T0222709

Data-Handling in Biomedical Science

Packed with worked examples and problems for you to try, this book will help to improve your confidence and skill in data-handling. The mathematical methods needed for problem-solving are described in the first part of the book, with chapters covering topics such as indices, graphs and logarithms. The following eight chapters explore data-handling using different areas of microbiology and biochemistry including microbial growth, enzymes and radioactivity as examples. Each chapter is fully illustrated with worked examples that provide a step-by-step guide to the solution of the most common problems. Over 30 exercises, ranging in difficulty and length, allow you to practise your skills and are accompanied by a full set of hints and solutions.

PETER WHITE taught practical classes and tutorials in microbiology for over 30 years during his time as a Lecturer and Senior Lecturer at Sheffield University. His research interests include microbial drug resistance, biochemistry of bacterial walls and metabolism of members of the genus *Bacillus*. He is a member of the Society for General Microbiology and a past member of the Biochemical Society and the American Society for Microbiology.

Data-Handling in Biomedical Science

PETER WHITE

University of Sheffield

CAMBRIDGE
UNIVERSITY PRESS

CAMBRIDGE
UNIVERSITY PRESS

Shaftesbury Road, Cambridge CB2 8EA, United Kingdom

One Liberty Plaza, 20th Floor, New York, NY 10006, USA

477 Williamstown Road, Port Melbourne, VIC 3207, Australia

314–321, 3rd Floor, Plot 3, Splendor Forum, Jasola District Centre, New Delhi – 110025, India

103 Penang Road, #05–06/07, Visioncrest Commercial, Singapore 238467

Cambridge University Press is part of Cambridge University Press & Assessment, a department of the University of Cambridge.

We share the University's mission to contribute to society through the pursuit of education, learning and research at the highest international levels of excellence.

www.cambridge.org
Information on this title: www.cambridge.org/9780521143868

© P. J. White 2010

First published 2010

A catalogue record for this publication is available from the British Library

Library of Congress Cataloging-in-Publication data
White, Peter, 1931–
Data-handling in biomedical science / Peter White.
    p. cm.
Includes index.
ISBN 978-0-521-19455-6 – ISBN 978-0-521-14386-8 (pbk.) 1. Microbiology – Data processing. I. Title.
QR69.D35W45 2010
616.9′041–dc22

                                                          2009053993

ISBN   978-0-521-19455-6   Hardback
ISBN   978-0-521-14386-8   Paperback

Dedication

To all my teachers, but especially to remember
Miss Dollan
(Leigh Church of England Junior School 1941–1942)

Contents

Introduction

Data-handling means the interpreting and refining of experimental results. This book is aimed at helping to improve confidence and skill in data-handling. It is intended for undergraduate students, and for graduate students who may still have a little to learn.

Although microbiology began with simple observations (the organisms are small, they have various shapes, and some are motile) the subject has become a quantitative, experimental science. As an example consider the following statement:

> Poly β-hydroxybutyrate may make up 70% of the dry weight of *Azotobacter*.

To reach this conclusion one must grow the organisms in a certain way, make weighings, extract the polymer, do a chemical assay, and then put all the information together in its proper order. None of these practical steps is difficult, but to achieve the final result, clarity of thought, rather than great mathematical ability, is definitely needed.

In recent times, the words 'data-handling' have taken on a second meaning, that is, the manipulation of very large quantities of data (such as DNA sequences) by using computer programs for analysis and comparison. This new big area of database management is not covered here. Everything in this book can be done with pencil and paper and a pocket calculator. That is by no means to decry computers; the whole of this text was written on a word processor, and all the figures have been drawn with Excel® or Corel Draw®.

How is data-handling to be learned? Best of all, **by personal experience in the laboratory**. If you design an experiment yourself, then you will have thought about the form of your results and the way in which they will need to be manipulated. If you do an experiment that someone else has designed (as in a practical class) it may be harder, but is still very instructive, to work out how to do the data-handling. When you are told how to do it, then make sure that you really understand the steps; this is where many students fall short.

The other way to learn is by solving written problems. Reading through problems, without seriously attempting to attack them, is a waste of time. Listening to someone explaining how a problem is done may be instructive, but it can make things seem too easy (like seeing the solution to a crossword puzzle) or it can make things appear too difficult (when really you could have done it if you had tried).

How is data-handling to be taught? The answer is simple: by encouraging **practice, practice and practice** with written problems and at the bench with real situations. People who are not alarmed by simple mathematics or who like puzzles might have an advantage in data-handling, yet practice will improve anyone's ability. Becoming familiar with types of calculation that occur again and again builds confidence to tackle new situations. Building this confidence is extremely important, and is best done by starting with easier problems and moving gradually towards the harder.

This book starts with four chapters about simple mathematics and one about graphs. All of this material can be skipped, but it has to be well understood before attempting the later chapters. The next two chapters, about logarithms and statistics, are more difficult. However, these should not be skipped unless you are very assured in both topics.

After this come eight chapters in which data-handling in different areas of microbiology and biochemistry are discussed and are illustrated by fully worked-out problems. Finally, a miscellany of problems is given, with the answers separated to a following chapter.

These problems were designed by various people; their names are given at the starts of the problems; no name means by me. All the answers and any errors are my sole responsibility. I hope the quotations may be seen to have some relevance, and not to be just show-offs. Blame Edgar Allan Poe or Colin Dexter.

How important is maths in data-handling?

The answer is vitally, but do not despair. There are few areas of microbiology or biochemistry that require any kind of advanced mathematical ability. What is necessary is to know how to crank out answers from standard methods. Most students, and most of their teachers too, have not been trained to a high level in mathematics. In the author's case this means trained only as far as School Certificate, the forerunner of 'O' level. An expert is usually consulted in those rare cases where advanced knowledge is needed.

Students are expected to be numerate, and modest skills are certainly necessary. Appreciation of simple proportion, knowledge of some easy algebra (such as solving simultaneous equations), an understanding of logarithms and basic statistics, and the ability to draw and interpret graphs are all needed. These topics are revised in the first chapters of this book. However, most important of all is to develop a confidence that nearly all data-handling problems can be tackled without having to be a talented mathematician.

Clear thinking and simple mathematics will solve most problems (at least in this branch of science)!

The fact is that there are few more 'popular' subjects than mathematics. Most people have some appreciation of mathematics, just as most people can enjoy a pleasant tune, and there are probably more people really interested in mathematics than in music. Appearances may suggest the contrary, but there are easy explanations. Music can be used to stimulate mass emotion, while mathematics cannot; and musical incapacity is recognized (no doubt rightly) as mildly discreditable, whereas most people are so frightened of the name of mathematics that they are ready, quite unaffectedly, to exaggerate their own mathematical stupidity.

G. H. Hardy (1940)

Abbreviations and the Système International

When the same word or phrase appears many times in a piece of writing, common practice is to abbreviate. For instance, in this book adenosine triphosphate is written as ATP; reduced nicotinamide adenine dinucleotide is NADH. Some abbreviations are understood by every reader of English, like Mr, USA, Prof., i.e. (= id est = that is). Other abbreviations, for example µg, are only understood by specialists, and others may be too new, such as cu (= see you), for everyone to understand.

In science there is a need for abbreviations that are intelligible to all readers in any language. In order to express ten grams one abbreviates as 10 g; to write twenty seconds one puts 20 s (note that numerals are abbreviations too); and one hundred degrees Centigrade becomes 100 °C. There may be times, of course, when you don't want to use an abbreviation, and then it might be appropriate to write (say) 100 millilitres rather than 100 ml.

The Système International (SI) is an agreement that in all science there shall be certain basic units for measuring distance (the metre, m), time (the second, s), weight (the kilogram, kg), temperature (degree Kelvin, K), luminous intensity (the candela, cd), electric current (the ampere, A), and quantity (the mole, mol). There are multiples or fractions of these units, some of which have names of their own (e.g. 1 min = 60 s; 1 h = 3600 s). Derived units are formed by combination of two or more of these basic units; the joule (J) is $m^2.kg.s^{-2}$ (= $m^2.kg.s^{-3}$).

SI units are now used in all scientific writing. **Their abbreviations are fixed and are never pluralised and never have a full stop following (unless at the end of a sentence or in place of a multiplication symbol [×]).** Thus, in an article for a journal you can write 50 g but cannot have 50 g. or 50 gs or 50 gm or 50 gms. Extensive lists of SI units are available in libraries and on the Internet with their definitions and proper abbreviations, and with quite elaborate rules too. It is important to know the abbreviations for units that you may frequently use.

The litre (1000 ml) is a special problem. The SI abbreviates litre as l (hence ml, μl etc.) and trouble comes when you want to express a number of whole litres, like 10 l, or 100 l, or to abbreviate 'per litre' (l^{-1}). In these cases, where l looks too like a 1 or an I, then I think it is better to write 10 L and 100 L, or L^{-1}. This usage is common, and I follow it. In the USA ml and μl will often be written as mL and μL, but I am not so thoroughly Americanised (or consistent) as to do this.

Any abbreviation that may not be understood by one's entire readership must be explained. If in any doubt then explain. The medical profession coins abbreviations that all too often go without explanation. What about MRSA? It is improbable that every reader knows that this contraction means 'methicillin-resistant *Staphylococcus aureus*' and not something like 'most repulsive slimy animal' that you meet in a hospital.

Abbreviations used in this book

SI units and abbreviations are used wherever possible. Note the litre anomaly (above).

The erg is the full name of a unit, not an abbreviation, and because it was defined in terms of $cm^2.g. s^{-2}$, the erg is equal to 1×10^{-7} J.

Prefixes

M   mega- ($\times 10^6$)
k   kilo- ($\times 10^3$)
m   milli- ($\times 10^{-3}$)
μ   micro- ($\times 10^{-6}$)
n   nano- ($\times 10^{-9}$)

Other abbreviations

A	adenine
ADP	adenosine diphosphate
Ala	alanine
ATP	adenosine triphosphate

C	cytosine
cal	calorie
CoA	coenzyme A
CoQ	coenzyme Q (ubiquinone)
Da	dalton = unified atomic mass unit [= 1 / (Avogadro's number) gram]
Dap	2,6-diaminopimelic acid
DEAE-	diethylaminoethyl-[cellulose]
DNA	deoxyribonucleic acid
e.g.	for example
F	the Faraday; the total electrical charge on 1 mole of electrons
FAD	flavin adenine dinucleotide (oxidised form)
$FADH_2$	flavin adenine dinucleotide (reduced form)
G	guanine
Glu	glutamate
i.e.	that is
Lys	lysine
mol. wt	molecular weight (= relative molecular mass, RMM)
M	molar (note use of smaller typeface compared with the prefix M = mega)
mur	muramic acid
NAD^+	nicotinamide adenine dinucleotide (oxidised form)
NADH	nicotinamide adenine dinucleotide (reduced form)
$NADP^+$	nicotinamide adenine dinucleotide phosphate (oxidised form)
NADPH	nicotinamide adenine dinucleotide phosphate (reduced form)
OD	optical density
P_i	inorganic phosphate
PP_i	inorganic pyrophosphate
PQQ	pyrroloquinoline quinone
R	the universal gas constant (1.986 calories per degree Kelvin per mole)
STP	standard temperature (273 degrees Kelvin) and pressure (1 atmosphere, 760 mm mercury)
T	thymine
T	temperature in degrees Kelvin
UDP	uridine diphosphate
t_d	doubling time during exponential growth

UV	ultra violet
V	volt
v.	versus (= against)
v/v	volume per volume
wt	weight
w/v	weight per volume

Some other abbreviations may be defined where they occur in the text.

Acknowledgements

I am grateful to all those students and staff, academic, technical and secretarial in the Department of Molecular Biology at Sheffield University who have helped so much in the making of this book. Particularly I want to thank Dr D. J. Gilmour and Professor Anne Moir for their interest and advice.

Martin Griffiths and Lynette Talbot, of Cambridge University Press, have been most friendly and helpful throughout production. It has also been a pleasure to work with my copy-editor and cricket-lover, Anna Hodson.

Most of all, of course, I thank my wife, Ruth, for just about everything.

1 | Numbers and indices

Most of this chapter should be familiar, but it is important that you really understand all of the material, which is largely a series of definitions.

1.1   Numbers

Real numbers are numbers that can be fitted into a place on the number scale (Fig. 1.1). The other kinds of numbers are **complex** (or **imaginary**) **numbers**, which cannot be fitted onto this scale, but lie above or below the line. They are of the general form $a + ib$, where a and b are real numbers but i is the square root of -1.

Real numbers can be divided into:

Integers: these are whole numbers, positive or negative, such as 7, 341, −56.

Rational numbers: these can be expressed *precisely* as the ratio of two integers. All integers are rational (they can be written as $n / 1$) and many non-integers are also rational, such as 3 / 4, 2.5 (= 5 / 2), −7.36 (= −736 / 100).

Irrational numbers: these *cannot be precisely* expressed as the ratio of two integers; examples are π (which is not exactly 22 / 7 nor any other ratio of integers) and the square roots of all prime numbers (except 1). Note that a number that has to be written as a recurring decimal is not irrational: 0.333 333 … is exactly 1 / 3; and 0.142 857 142 857 142 857 … is 1 / 7. Also, all approximations are rational: if we give π the approximate value of 3.142 this is 3142 / 1000, a rational number.

Fig. 1.1 The scale of real numbers.

1.2   Indices

A number written as n^a is defined as **the number n raised to the power a**. If a is a positive integer (the simplest case) then n^a means that n is multiplied a times by itself (n). Thus, 2^5 means $2 \times 2 \times 2 \times 2 \times 2 = 32$. In the expression n^a the number n is called the **base**, and a is called the **index** (or power, or exponent). Neither n nor a need to be integers.

Expressions that include more than one base cannot always be simplified: nothing for instance can be done with an expression such as $n^a \times z^b$. However, if the power is the same, we may be able to rewrite, as for example, $n^a \times z^a = (nz)^a$. Simplifications are possible in some very important cases where only one base is present, as follows.

Multiplication

$$n^a \times n^b = n^{(a+b)}$$

e.g. $5^2 \times 5^3 = (5 \times 5) \times (5 \times 5 \times 5) = 5^5 = 25 \times 125 = 3125$

Note carefully that this only works when the base is constant. Expressions such as $n^a \times z^b$ cannot be treated in this way.

Division

$$n^a / n^b = n^{(a-b)}$$

e.g. $2^4 / 2^3 = (2 \times 2 \times 2 \times 2)/(2 \times 2 \times 2) = 2^1 = 2$

Again the base must be constant for this to work.

These two relationships are the foundation for the use of logarithms (which are themselves indices of a chosen base number) as aids to multiplication and division.

Powers of indices

$$(n^a)^b = n^{a \times b}$$

e.g. $(3^2)^3 = 3^2 \times 3^2 \times 3^2 = (3 \times 3) \times (3 \times 3) \times (3 \times 3) = 3^6 = 729$

Note carefully that expressions such as $n^a + n^b$ or $n^a - n^b$ cannot be simplified (unless the actual numerical values of n, a and b are known).

e.g. $3^3 + 3^1 = (3 \times 3 \times 3) + 3 = 27 + 3$

$\qquad = 30 \quad$ and not 3^4 (which equals 81),

$3^3 - 3^2 = (3 \times 3 \times 3) - (3 \times 3)$

$\qquad = 27 - 9 = 18 \quad$ and not 3^1 (which equals 3)

Indices need not be only positive integers. They may also have zero value (e.g. n^0), or be negative (e.g. n^{-3}) or fractional (e.g. $n^{3/2}$). It is important to understand what these different usages mean.

Any base raised to the power 0 has a value of 1:

$$n^0 = 1$$

$$(n^a \div n^a = 1 = n^{(a-a)} = n^0)$$

Any base raised to the power 1 has a value equal to the base itself:

$$n^1 = n$$

A base raised to the power 0.5 has a value equal to the square root of the base:

$$n^{0.5} \times n^{0.5} = n^1$$

A base raised to a negative power represents the reciprocal of the base raised to that same (but positive) power:

$$n^{-a} = 1/n^a$$

The value of $n^{a/b}$ is the bth root of n^a:

$$n^{a/b} = \sqrt[b]{n^a}$$

So, for example, $2^{5/4} = \sqrt[4]{2^5} = 2^{1.25}$ and $3^{-3/2} = 1/\sqrt[2]{3^3} = 1/3^{1.5}$. These kinds of expression are most easily solved by using logarithms (or a pocket calculator), as will be discussed later (see Chapter 5).

Standard form

In order to write big (or small) numbers in a compact way we express them as powers of 10, for example:

$$234\,700\,000 = 2.347 \times 10^8; \quad 0.000\,000\,625 = 6.25 \times 10^{-7}$$

While these numbers could just as well be written as 23.47×10^7 and 0.625×10^{-6}, the standard form is to show a single integer (other than 0) to the left of the decimal point.

Two things to be careful about:

(1) Going from standard form to a written-out number can be treacherous and so take **great** care. 2×10^{-3} does not equal 0.02: rather, 2×10^{-3} $= 0.002$. This seems obvious yet this kind of error is common.

(2) You can add or subtract numbers in standard form **only** when all the numbers are rewritten each at the same power of 10. At the end you can convert back to standard form if necessary.

$$(3 \times 10^3) + (8 + 10^2) - (5 \times 10^1) = (300 \times 10^1) + (80 \times 10^1)$$
$$-(5 \times 10^1)$$
$$= 375 \times 10^1$$
$$= 3.75 \times 10^3$$

Getting this right looks easy but is really quite troublesome. An example on the Internet that shows how to do this kind of calculation is worked out to the wrong answer! The safest thing is to write out all the numbers fully (i.e. as multiplied by 10^0):

$$3000 + 800 - 50 = 3750 = 3.75 \times 10^3$$

Multiplying (or dividing) numbers in standard form is relatively easy:

$$(3 \times 10^3) \times (8 \times 10^2) \times (5 \times 10^1) = 3 \times 8 \times 5 \times 10^{(3+2+1)}$$
$$= 120 \times 10^6$$
$$= 1.20 \times 10^8$$
$$(8 \times 10^2) \div (5 \times 10^1) = (8 \div 5) \times 10^{(2-1)} = 1.6 \times 10^1$$

2 | A sense of proportion

If the Eiffel tower were now representing the world's age, the skin of paint on the pinnacle-knob at its summit would represent man's share of that age, and anybody would perceive that the skin was what the tower was built for. I reckon they would, I dunno.

Mark Twain

The object of this chapter is to encourage you to think whether or not your answer to a problem looks reasonable or ridiculous. In general, a reasonable answer is likely to be a right answer. An answer that looks ridiculous might also be right, but you should then be alert to check your calculation very carefully. Of course, there will be times when you do not know what to make of an answer – is it reasonable or is it not? The better your background of knowledge and experience, the less often will this uncertainty happen.

2.1   A ridiculous answer that is wrong

Here is the problem: calculate what dry weight of bacteria will be present in 10 litres of medium in a fermenter after 10 h when at time zero there are 10 organisms ml^{-1} and there is a lag of 1 h before exponential growth (doubling time 20 min) begins. One organism has a dry weight of 1×10^{-12} g.

This is the answer from a candidate in an examination (examiner's comments in [ ]):

There are 9 hours of exponential growth

In 1 hour there are 3 doublings ($t_d = 20$ min)

Therefore there are 27 doublings in total

So that 10×2^{27} organisms will be present per ml after 10 hours [*perfect so far*]

$= 2 \times 10^{27}$ organisms per ml [*spectacularly wrong; needs to read about indices*]

$\equiv 2 \times 10^{27} \times 1 \times 10^{-12} \times 10^{4}$ g dry weight in 10 litres [*right if line above were right*]

$= 2 \times 10^{19}$ g

Now the candidate starts to worry. [*The sadistic examiner begins to be amused*]

$= 2 \times 10^{16}$ kg

[*The examiner is laughing*]

$= 2 \times 10^{13}$ metric tonnes

This would not fit in the fermenter, *writes the candidate as the last line of answer.*

[*The examiner is rolling on the floor; marking scripts has its compensations*]

This answer is plainly ridiculous, but it is not clear whether the candidate has realised there is a mistake in the calculation, or (more probably) whether the examiner is being implicitly criticised for setting a problem that has a stupid answer. As this latter circumstance never happens (well, hardly ever), then there must be a mistake, as is pointed out above:

10×2^{27} does not equal 2×10^{27}. Rather, $10 \times 2^{27} = 10 \times 1.342 \times 10^{8}$

The correct dry weight after 10 hours is $1.342 \times 10^{9} \times 1 \times 10^{-12}$ g ml^{-1} = 1.342×10^{-3} g ml^{-1} or 1.342 mg ml^{-1} and so 13.42 g in 10 L

This answer does not appear impossible, and looks plausible if one has some knowledge of the levels of growth that bacterial cultures typically reach (1 to 10 mg dry weight ml^{-1}).

Simple mistakes in calculation are the commonest reason for getting wrong answers. Always think about the likely size of a result, and be sure to get ratios the right way round. For example, if you are finding how much of an anhydrous compound to use in a solution, when the recipe calls for a hydrated salt, then the required amount will be smaller than the recipe says. Remembering that where x is a positive real number:

multiplying x by a positive number less than 1 will lead to a number smaller than x

multiplying x by a number bigger than 1 will lead to a number bigger than x

dividing x by a positive number less than 1 will lead to a number bigger than x

dividing x by a number bigger than 1 will lead to a number smaller than x

should help you to express proportions the correct way round.

2.2   'Back of envelope' calculations

One of the most useful things you may learn from this book is how to get an approximate idea of the answer to a calculation by doing quite drastic rounding up and rounding down of the numbers in an expression. For example:

2875 × 7681 can be rounded to 3000 × 7500 which is 22 500 000
(If one number is rounded up, try to round down another.)
The precise answer is 22 082 875, and so the approximation is only off
   by 1.9%.

Here is another: (3478 × 29 641) / (391 × 475) can be written as:

$$\frac{3478 \times 29\,641}{391 \times 475} \longrightarrow \frac{3500 \times 30\,000}{400 \times 500}$$

$$\frac{\overset{175}{\cancel{700}}\;\cancel{3500} \times \cancel{30\,000}}{\underset{1}{\cancel{400}} \times \underset{1}{\cancel{500}}} \longrightarrow 525$$

Rounding the numbers allows drastic cancelling, to give an answer only 5.4% away from the precise result, which is 555.1.

Even if you can do this more quickly with a calculator, you can also easily make mistakes in pressing wrong keys, and for many people disbelieving what the calculator says is difficult. The last example (below) is none too simple with a calculator, and back-of-envelope work is highly desirable to get an idea of what to expect as the answer.

$$\frac{(7836 - 484) \times 9741}{(2743 \times 37) + 960} \longrightarrow \frac{(8000 - 500) \times 10\,000}{(3000 \times 40) + 1000}$$

$$\longrightarrow \frac{7500 \times 10\,000}{120\,000 + 1000} \longrightarrow \frac{7500 \times 10\,000}{120\,000}$$

$$\longrightarrow \frac{\overset{2500}{\cancel{7500}} \times \cancel{10\,000}}{\underset{4}{\cancel{120\,000}}} \longrightarrow 625$$

The precise answer is 699.0, which means that this time the rough answer is not so close; the error is 10.6%, but even so this still gives a good idea of what to expect as the correct result after accurate calculation.

You probably think that all these examples were carefully devised, with the roundings planned ahead to get a good answer. They were in fact done with no forethought of that kind, and are genuine, honestly. Doing back-of-envelope calculations without an envelope (i.e. in your head) is a talent that can be shown off to the uninitiated (impress your friends!), but be very careful not to lose track of powers of 10. The envelope is safer.

3 | Graphs

'and what is the use of a book,' thought Alice, 'without pictures or conversation?'

Lewis Carroll

Why draw a graph? There are many reasons, but the fundamental one is that the human brain understands a picture much more easily than it does a table of numbers.

Many data-handling questions require a graph to be drawn as part of the solution. It is unlikely that under examination conditions a work of art will be produced, nor would one be expected. However, some marks are given for a graph that is correct (the points are plotted in the right places!) and which obeys the conventional rules.

As well as making the drawing, you will probably have to use the graph to read off some values, such as a gradient or an intercept or to measure test samples from an assay. Doing these interpretations will be considered after discussing how to produce a graph.

3.1 Drawing graphs

The graph shown in Fig. 3.1 illustrates a number of features.

There are several things to note. The horizontal scale (x axis, or abscissa) is given to the variable that is the more directly under the control of the investigator, and the variable that is measured for various values of x is plotted on the vertical scale (y axis, or ordinate). In Fig. 3.1, the times at which readings of the extinction are made are chosen by the experimenter, and so go on the x axis, while the extinctions themselves are less under control and follow from the selected times, and therefore go on the y axis.

Do not make the graph too small; aim to use as much of the area of the sheet of graph paper as possible. The scales of the two axes must therefore be chosen with care. Neither scale should extend far beyond the plotted points,

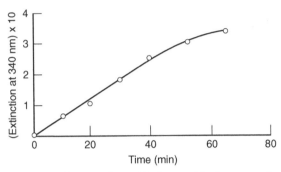

Fig. 3.1 Reduction of NAD^+ by lactate dehydrogenase. At time zero, enzyme, substrate and cofactor were mixed and extinctions at 340 nm (due to formation of NADH) were measured at 30 °C at intervals. Readings were continued for 66 min from the time of mixing, and the measured extinctions fell in the range 0 to 0.32.

and the points themselves must always lie within the scales. Straight-line graphs are best plotted with scales devised in such a way that the line makes an angle of about 45° with the x axis, so that x or y values can be plotted or read from the graph with similar precision. (When several lines are plotted on one graph it is unlikely that this rule can be obeyed for all the lines, but at least one of the lines ought to be plotted to best advantage.)

The origin of a graph does not necessarily have to be shown. Frequently a better graph can be made by using scales of limited ranges. Compare Fig. 3.2a and b.

A problem sometimes occurs in the laboratory with real experimental results: '*Must* a straight line be drawn to go through the origin even though doing this gives a line of poorer fit with the data?' Unfortunately, the answer is sometimes 'Yes' and sometimes 'No' depending on all kinds of things. Fortunately the made-up results given in a data-handling question will not be equivocal (unless you are specially warned!) and will not lead to points on graphs that leave much doubt about where the curve ought to go, that is, provided the points are plotted in the right places. Real-life problems may not be so amenable!

Frequently points are not plotted correctly. People often miscount squares on graph paper and hence make scales with irregular spacing of the scale divisions. Another common error is to choose a logarithmic scale when a linear one ought to have been used. If you have numbers spread between 10 and 10 000 to plot (e.g. organisms per ml), and you label the

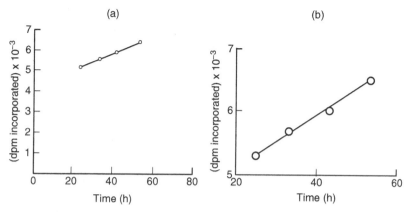

Fig. 3.2 Scales of graphs need not always include the origin. The same data are here plotted on two different scales. The graph on the right (b) gives a more accurate representation of the results. Note incidentally that the scales on both graphs are linear (0, 1000, 2000, 3000 etc. and 5000, 6000, 7000) and not logarithmic.

divisions of the y axis as 0, 10, 100, 1000, 10 000 organisms ml^{-1} (with equal spacing between each) then you have created a logarithmic scale (with base 10). To use this scale properly a result of 200 must be plotted at 0.301 of the interval between 100 and 1000, not at 0.2 of the interval. Likewise, 500 should be at 0.699 of the interval between 100 and 1000, not at 0.5 of the interval. This kind of mistake can often be seen in published papers. A linear scale for the same numbers (0 to 10 000) would be 0, 2000, 4000, 6000, 8000, 10 000 with equal spacing between each. Values less than 100 will look very small, and even 5000 will be only halfway up the y axis.

Ordinary graph paper is marked out as 1-cm squares each divided horizontally and vertically by lines 1 mm apart. As a consequence, it is easy to plot points correctly when each 1-cm square is made to represent 1 unit or 10 units or 0.1 unit, etc. Having each 1-cm square represent 2 or 5 units is manageable, but a bit more difficult. Choosing any other number of units per 1-cm square is an excellent recipe for errors.

Each axis must bear an explanation of what numbers (and their units) are plotted. In Fig. 3.1 it is obvious that the x axis should be labelled 'Time (min)'. The y axis (in this instance) is more difficult to get right, and it gives an example of avoiding a very common kind of mistake. The numbers that have to be plotted are extinctions of 0 to 0.32. Whenever possible, it is best to use small whole numbers to indicate the scale, and so the actual measured extinctions have been **multiplied by 10^{+1}** before being plotted as 1, 2, 3, etc. and **not by 10^{-1}** (which would mean that the measured extinctions were 10, 20, 30, etc.). Many people find this hard to understand, and at first it may seem the opposite of common sense, though in truth it is not. Here are some examples to reinforce the argument. If units of 1000, 2000, 3000 are plotted as 1, 2, 3 then the scale should be labelled 'units $\times 10^{-3}$' and not 'units $\times 10^{3}$' (which would mean that the actual values were 0.001, 0.002, 0.003). If units of 0.01, 0.02, 0.03 are plotted as 1, 2, 3 then the scale should be labelled as 'units $\times 10^{2}$' and not as 'units $\times 10^{-2}$' (which would mean that the actual values were 100, 200, 300).

The index marks on the x and y scales should extend **into** the area of the graph, and not outwards towards the scale. Unfortunately many computer programs draw these marks in the wrong direction. You can correct this before printing, if you take the trouble.

Plotted points should normally be symbols (e. g. ● ○ ■ □) that are big enough for the reader to see them easily. If only one curve is plotted on a graph then all the points should be indicated by the same symbol. If more than one curve is plotted then points for each separate curve *must* be indicated with different symbols.

How best to connect the plotted points with a curve can be difficult to judge. Most often it is preferable to draw a smooth curve of best fit rather than to join points one by one with short straight lines. If a single straight line fits reasonably through several successive points then draw it. **Linear regression** is a method of calculating the straight line of best fit through a series of experimental values that are believed to be in a linear relation to one another. Many pocket calculators can do this useful analysis. However, beware of curves like Fig. 3.1 that are linear over only a limited range of values. This is a common shape for graphs showing the progress of a reaction or microbial growth, or for standard curves in spectrophotometric assays (see Fig. 3.3).

Few people can draw curved lines elegantly by freehand. Best results come after considerable practice with a set of French curves, but this method is slow and impractical (even if you are good at it) in an examination. Using a stiff, bendable ruler ('Flexicurve'™) is the most convenient expedient.

Do not extend curves beyond the plotted points unless this is done for legitimate extrapolation.

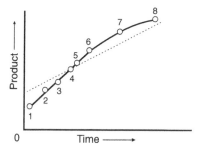

Fig. 3.3 The course of an enzymic reaction. The solid curve shows the correct relation between the points 1 to 8, i.e. a straight line of uniform gradient from point 1 to point 6, and then a curve of decreasing gradient from point 6 to point 8. The dotted line is the best-fit line (calculated by linear regression) when all the points 1 to 8 are regarded as being in a straight line relation. The gradient of the dotted line gives a considerable underestimate of the true initial rate of reaction.

Avoid plotting more than three curves on one graph, because the result usually then becomes too hard to understand readily. On the other hand, it can be helpful to plot two or three separate measured properties of a system (particularly plotted against time) on the same graph, as this may demonstrate unexpected relations between the different properties.

3.2 Reading values from graphs

Intercepts and gradients

Often it is necessary to determine one or both of these features of a curve. Finding the intercept (on x or y axis) is easy, though extrapolation is usually needed (Fig. 3.4). Finding the gradient of a straight line is also simple (see Fig. 3.5), but the gradient of a curved line changes along its length and so the value of the gradient depends on where it is measured (see Fig. 3.6). (It is very unlikely that you would be asked to find the gradient of a curved line in an examination.)

If you know the equation that defines a curve (e.g. $y = 2x + 5$; $y = x^2 - 3$) then intercepts (the value of y when $x = 0$ or of x when $y = 0$) and gradients can usually be found more easily by calculation than by drawing a graph (see Chapter 4).

In Fig. 3.5 the gradient is positive, but if the line had sloped down from left to right then the gradient would have been negative (y decreasing as

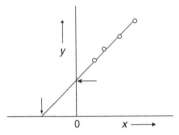

Fig. 3.4 Extrapolation from the plotted points to find the intercepts of a straight line on the y axis (a positive value in this example) and on the x axis (a negative value here).

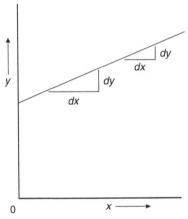

Fig. 3.5 The gradient of a straight line is the change in y for a given increase of x. The units of the gradient are the units of y divided by the units of x.

x increased). Notice that the gradient can be measured anywhere along the line and that taking a large or a small increment of x will yield the same value for the gradient (though the large increment will have greater accuracy).

Reading off test samples from a standard curve

This is the simplest application of a graph. On the standard curve one plots (as x values) known amounts of the authentic reference substance and the responses, which may be extinction values, number of organisms growing, etc. are plotted on the y axis. The response to a test sample is measured, and

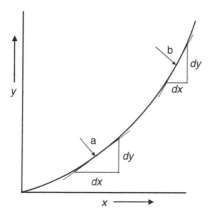

Fig. 3.6 The gradient of a curve. The gradient at point a is defined as the gradient of the tangent to the curve at point a. At point b the gradient of the tangent is obviously greater than at point a.

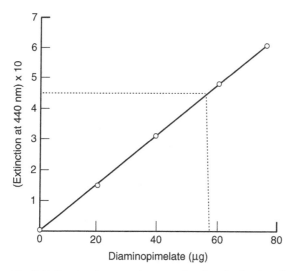

Fig. 3.7 Spectrophotometric assay of authentic diaminopimelic acid and a test sample (0.5 ml). (Incidentally, this assay is the most reproducible and easy colorimetric method that the author has encountered. It was devised by Dr Elizabeth Work.)

from this y value a corresponding amount of reference substance present in the assay system is read from the x axis (Fig. 3.7). Standard amounts of diamino-pimelate established the x and y axes. The extinction (0.45 in this example) due to the test sample is converted into an amount of diaminopimelate (58 μg).

This amount was present in the volume of test sample that was assayed, so that the sample contained 116 µg diaminopimelate ml^{-1}.

Finding doubling times of exponentially growing microbial cultures and half-lives of radioisotopes from plots of logarithms

These procedures are described in Chapter 5.

3.3 Solving equations graphically

Sometimes a difficulty in a calculation is most easily solved by drawing a graph. Consider the following problem:

In an aerated liquid medium at 37 °C bacteria of strain A grow exponentially $(t_d = 45$ min) after a lag phase of 3 h. Under the same conditions bacteria of strain B grow exponentially $(t_d = 30$ min) after a lag phase of 2 h. The medium is inoculated at the same time with 1000 organisms of strain A and with 20 organisms of strain B and then incubated. Determine the time (after inoculation) at which equal numbers of organisms of each strain are present, and find what is the total number of bacteria present at this time. Assume that neither strain affects the growth of the other, and that exponential growth of both strains begins immediately after the lag phase. Nutrients in the medium are sufficient to allow exponential growth to go on after the time when the numbers of each strain were equal.

Solving this problem by calculation alone is possible, but it is not at once obvious how to start, and it is a bit difficult to avoid mistakes, even when following the right steps. In contrast, a graph leads to the answers without much trouble. All that is necessary is to plot two straight lines representing the exponential growth of each strain, and find where they intercept; this is the time (x axis) at which the logarithms of the numbers of strains A and B are equal (y axis).

To plot exponential growth of strain A we need ln n_0 (that is ln 1000 at 180 min) and ln n_t where t can be any time greater than 180 min. If we arbitrarily make $t = 580$ min then there will be 400 min of exponential growth of this strain and so:

$$\ln n_{580} = \ln 1000 + \mu.400 \text{ (where } \mu = 0.693/t_d)$$
$$(\mu = 0.693/45 = 0.0154 \text{ min}^{-1})$$

(see Chapter 5 about logarithms and exponential growth of microorganisms). Thus,

$$\ln n_{580} = 6.9078 + (0.0154 \times 400) = 13.068$$

Now we can draw a line on a graph of $\ln n$ (y axis) against time (x axis) to join 6.9078 (y axis) at 180 min (x axis) with 13.068 (y axis) at 580 min (x axis). This line can be extended, if necessary, beyond 580 min because we are

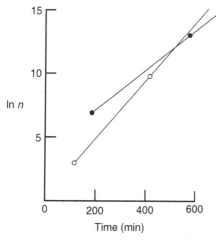

Fig. 3.8 Growth of two strains of bacteria as a mixed culture: ● strain A, ○ strain B; n is the number of organisms per ml.

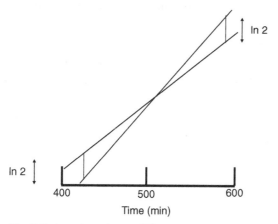

Fig. 3.9 An enlarged part of Fig. 3.8.

told that exponential growth continues after the time when numbers of strains A and B are equal.

In the same way, we can plot exponential growth of strain B. Here n_0 is 20 at 120 min, and μ is 0.0231 min^{-1}. If we consider 300 min of exponential growth of strain B then we arrive at:

$$\ln n_{420} = 2.9957 + (0.0231 \times 300) = 9.926$$

Plotting the values for strain B leads to the result shown in Fig. 3.8.

We can read off that the required time is 510 min, and the natural logarithm of the number of organisms of either strain at this time is 11.95, and so the number is $e^{11.95} = 1.55 \times 10^5$. Thus the total number of organisms is 3.10×10^5. The more precise answers given by calculation are 508 min and 3.12×10^5 organisms. Can you do the calculation? It's not very simple. A solution is in Chapter 17 (Answers to problems).

Suppose the problem also asked you to establish the times between which neither strain is in more than a twofold excess over the other. This again can be calculated, but the answer can easily be read from the graph already drawn. All that is needed is to find the region where the two lines are not more than 0.693 (on the ln n scale) apart, since this distance represents a factor of 2 (ln 2 = 0.693). An enlarged part of the graph is given in Fig. 3.9, and it shows that the required period is between 420 min and 590 min. Calculation gives the more precise 418 min and 598 min (see Answers to problems).

A bit of geometry (in biology?) shows that the two triangles in Fig. 3.9 are congruent (the vertical lines are parallel and equal in length), which can prove that the period of time, before the point of equal numbers, when strain A is in not more than twofold excess, must be numerically the same as the period after the point of equal numbers when strain B is in not more than twofold excess. I haven't thought about how to *prove* this without the graph.

See Chapter 4 (Algebra) for more examples of the use of graphs.

See Chapter 5 (Logarithms) if you don't understand e and natural logarithms (ln n).

4 | Algebra

I would advise you, Sir, to study algebra, if you are not already an adept in it: your head would get less muddy.

Samuel Johnson

Biologists use simple algebra to devise and solve equations. All we shall show in this chapter is how to do these things:

Devising equations
Substituting into equations
Solving equations: simultaneous and quadratic
Rearranging equations.

4.1 Devising equations

Let's start with an easy example of devising an equation, treated at extravagant length.

A common problem is to find what volume of a bacterial suspension of known optical density (say 1.2) must be added to a tube containing a known volume of sterile water (say 5 ml) to give a dilute suspension of a desired lower optical density (say 0.1).

Answer:

Let $x =$ volume (in ml) of suspension (optical density 1.2)
that is required.

In order to get the correct dilute suspension, the ratio of x to the final total volume of diluted suspension $(5 + x)$ must be 0.1 / 1.2. So now we can write an equation:

$$x/(5 + x) = 0.1/1.2$$

Now solve for x. Multiply both sides of the equation by 1.2, which gives

$$1.2x/(5 + x) = 0.1$$

Now multiply both sides of the equation by $(5 + x)$, which gives $1.2\,x = 0.1 \times$ $(5 + x)$ which is $0.5 + 0.1\,x$, and so (by subtracting $0.1\,x$ from each sides of the equation) $1.1\,x = 0.5$. Thus (by dividing both sides by 1.1):

$$x = 0.5/1.1 = 0.45\,\text{ml}$$

To check the answer, show whether $0.45 / (5 + 0.45)$ does indeed equal $0.1 / 1.2$:

$$0.45/(5 + 0.45) = 0.083$$
$$0.1/1.2 = 0.083$$

Good, 0.45 ml is the right answer.

Note particularly the following points about these steps.

(1) When x (an unknown value) is created **it is essential to define fully the units of x** ('ml of a suspension of optical density 1.2' in the above instance). Failure to define units properly is a very common cause of mistakes: you need to be quite clear from the start that x is not ml of water, or ml of diluted suspension. In the same way, it is not enough (for instance) to define t as time in hours, you must also say time in hours from the start of incubation, or from the start of exponential growth, or from whatever point is appropriate.

(2) Setting up the equation to solve for x is the hardest part of the process to think out, and to explain as well. Trial and error and perhaps intuition are always needed. In the above example it is relatively easy to produce the equation. The ratio of x (volume of suspension wanted) to total volume $(x + 5)$ after adding to water must obviously be less than 1, and also must be equal to the ratio of the final to the initial optical density, which ratio is less than 1 too. As long as you remember that **if one side of an equation has a certain magnitude then the other side must be of the same magnitude** (because it is an *equation*) then you will avoid the mistake of writing $x/(5 + x) = 1.2 / 0.1$. I hope that this kind of error seems ridiculous. It often happens though.

(3) Solving our equation ought to need no more explanation. Answers don't come out more easily than this one.

4.2   Substituting into equations

The next example is a case where an equation does not have to be devised (though one does have to be remembered), but numbers must be inserted correctly into the equation to produce the right answer.

A problem that I had in the laboratory some time ago was to find the internal diameter (bore) of a uniform capillary tube of circular cross-section. A piece of the tube was first weighed empty. Then water (density 1.00 g ml^{-1}) was drawn into the tube, the sides of which were dried. The presence of the column of water (3.0 cm long) in the tube increased the weight by 0.0020g. What was the bore of the tube?

Answer: The water weighed 2 mg and so its volume was 2 µl (1 g of water occupies 1 ml, which is the same as 1000 mm^3 and 1000 µl).

The volume of a circular column is given by:

$$\text{volume} = \pi r^2 l$$

where r is the radius of the column and l is its length. Here it is vital that volume, r and l are expressed in compatible units. As the unit of volume is µl, r and l must be expressed in mm.

$$\text{Thus, } 2 = \pi r^2 \times 30$$
$$\text{Therefore } r^2 = 2 / (30\pi) = 0.0212 \text{ mm}^2$$
$$\text{and so } r = \sqrt{0.0212} = 0.146 \text{ mm}$$
$$\text{The bore of the tube is } 2r = 0.29 \text{ mm}$$

There are two slightly tricky points: one is to get the units right, and the other is to remember at the end that it is $2r$ (the diameter) and not just r that must be found.

4.3   Solving equations

The general procedure is to get the unknown (x) alone on one side of the equation and to have what x equals on the other.

In this next example x is on one side only, but in an awkward form which has to be simplified:

$$\frac{1}{x} + \frac{2}{3x} = 5, \text{ multiply each side by } x:$$

$$1 + \frac{2}{3} = 5x, \text{ rearrange}$$

$$5x = \frac{5}{3}, \text{ divide by 5}$$

$$x = \frac{1}{3}$$

Graphs and equations

Equations of the form $y = f(x)$, that is to say y is a function of x (e.g. $y = 3x - 2$; $y = 2x^3$; $y = x^2 + 2x + 3$) can be represented as graphs. One plots the values of y that are the result when values of x are put into the equation.

For example, the equation $y = 2x + 3$ gives the data:

x	y
-2	-1
-1	1
0	3
1	5
2	7

This information lets us draw a graph (Fig. 4.1). Any equation of the general form $y = mx + c$ (where m and c are real numbers) will produce a straight-line graph of gradient m and intercept c on the y axis (i.e. the value of y when $x = 0$). Intercepts on the y axis of other equations may be found by setting x to zero, but gradients of curves must be calculated by the mathematical process of differentiation, which is too big a subject for discussion here.

Quadratic equations

When one power of x alone is present in an equation there are no difficulties in solving for x, e.g. $x^4 = 27.98$, so that $x = \sqrt[4]{27.98} = 2.3$ or -2.3. (Remember that if $y = \sqrt[n]{x}$ and n is an even number then y will have both a positive and a negative value.) Things are less simple when there are more powers of x than just one. A quadratic equation is of the general form:

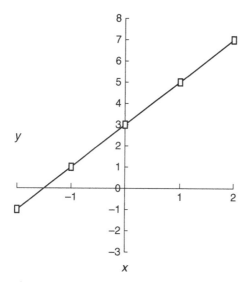

Fig. 4.1 Graph of $y = 2x + 3$. The straight-line graph has a gradient of 2, and intercepts the y axis at 3.

$$y = ax^2 + bx + c$$

where a, b and c are real numbers (positive or negative) that are known but which differ as one goes from one quadratic equation to another. (Equations with mixtures of more than two powers of x (polynomial equations) are too difficult for consideration here.)

Not all quadratic equations can be solved with real (as opposed to imaginary) numbers. Put the equation into the form $ax^2 + bx + c = 0$ (i.e. make y equal zero). To get the value(s) of x, substitute in the formula

$$x = [-b \pm \sqrt{(b^2 - 4ac)}]/2a$$

Then if (*and only if*) $b^2 - 4ac$ is positive or zero can the equation be solved for x. Thus, the equation $y = 3x^2 + 2x - 7 = 0$ is solvable because $a = +3$, $b = +2$ and $c = -7$, which means that $b^2 - 4ac = 4 - (4 \times 3 \times -7) = 4 + 84 = 88$, a positive value. However, $y = 3x^2 + 2x + 7 = 0$ is not solvable because $b^2 - 4ac = 4 - (4 \times 3 \times 7) = -80$, a negative value, for which no real square root exists.

For the equation $3x^2 + 2x - 7 = 0$, $x = (-2 \pm \sqrt{88}) / 6$ (see above)

Hence, $x = (-2 + 9.38) / 6$ or $x = (-2 - 9.38) / 6$

     $x = 7.38 / 6$ or $x = -11.38/6$

     $x = 1.23$ or $x = -1.90$

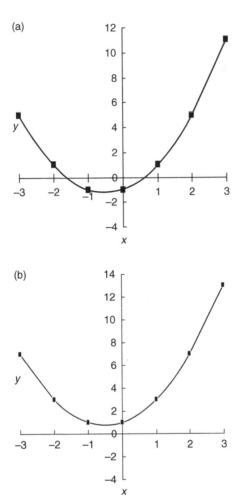

Fig. 4.2 Graphs of (a) $y = x^2 + x - 1$; (b) $y = x^2 + x + 1$; (c) $y = x^2 + 2x + 1$.

To check that these answers are correct, substitute back in the original equation:

$$(3 \times 1.23^2) + (2 \times 1.23) - 7 = 4.54 + 2.46 - 7 = 7 - 7 = 0$$
$$(3 \times -1.90^2) + (2 \times -1.90) - 7 = 10.83 - 3.80 - 7 = 0.03 \approx 0$$

Notice that two values for x are generally found from a solvable quadratic equation, but when $b^2 - 4ac$ is zero then only one value for x will be obtained. The curve of a non-solvable quadratic never cuts or touches the x axis (Fig. 4.2b), whereas the curve for a solvable equation usually cuts

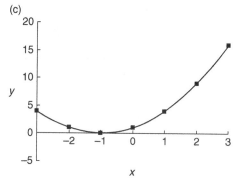

Fig. 4.2 (cont.)

the x axis twice (Fig. 4.2a) at the two values that are solutions of the equation. Only one solution is found when the curve touches the x axis in only one place (Fig. 4.2c). Find which of these equations are solvable for x by determining whether or not $b^2 - 4ac$ is positive:

$$x^2 = x + 1$$
$$x^2 = x - 1$$
$$x^2 = 1 - x$$
$$x^2 = 2x - 1$$
$$3x^2 / 4 = x + 1$$
$$8x = 3x^2 + 5$$
$$4 - x^2 = 5x$$

Now solve for x in each equation where this is possible.

(No answers are given for these little problems. You can check your answers by substituting them for x and seeing whether the equations balance.)

Simultaneous equations

A single equation that contains two unknowns (x and y) cannot be solved for x or for y. However, solution becomes possible when a second equation is available which relates the same numerical values of x and y (as in the first equation) to each other in a way different from the first equation.

Thus $3x + 2y = 17$ is not solvable alone, but if we also know that $x - y = -6$ then we can solve.

There is more than one way of getting to the answer; two non-graphical methods (a and b) are shown first, and then a graphical method (c).

(a) Make the multipliers of x (or of y) the same in each equation: thus, $x - y = -6$, so that $3x - 3y = -18$. Now subtract one modified equation from the other:

$$(3x + 2y = 17) - (3x - 3y = -18)$$

This leads to $5y = 35$, and so $y = 7$. To find x we put the established value of y into either of the two original equations:

$$3x + 2 \times 7 = 17 \text{ or } x - 7 = -6$$

and in each case we get $x = 1$.

(b) Express y in terms of x (or x in terms of y) by using one of the equations: $y = x + 6$. Now substitute for y in the other equation:

$$3x + 2(x + 6) = 17$$

This leads to $5x + 12 = 17$, and so $5x = 5$ and $x = 1$. To find y we put the established value of x into either of the two original equations:

$$3 + 2y = 17 \text{ or } 1 - y = -6$$

and we get $y = 7$ each time. In both procedures we use one equation to allow us to eliminate an unknown from the second equation, to leave a new equation with a single unknown.

(c) Put both equations into the form $y = f(x)$, thus, $y = (17 - 3x) / 2$ and $y = x + 6$. Now plot each of these equations as graphs (Fig. 4.3). The coordinates of the point of intersection of the two curves are the values of x and y that satisfy both equations.

Simultaneous equations cannot always be solved easily. Consider this pair of equations

$$2x + y = 22$$
$$xy = 36$$

Rearranging the first equation gives

$$y = 22 - 2x$$

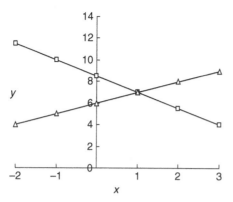

Fig. 4.3 Graph of $y = (17 - 3x)/2$ ($\square$) and $y = x + 6$ ($\triangle$). At the intersection $x = 1$ and $y = 7$.

so that

$$x(22 - 2x) = 36 = 22x - 2x^2$$

Rearranging again gives $2x^2 - 22x + 36 = 0$, which we can divide by 2:

$$x^2 - 11x + 18 = 0$$

Thus, we have a quadratic equation to solve in order to find x:

$$x = [11 \pm \sqrt{(121 - 4 \times 1 \times 18)}]/2$$
$$x = (11 + 7)/2 \text{ or } x = (11 - 7)/2$$
$$x = 9 \text{ or } x = 2$$

If x equals 9 then y must equal 4, and if x equals 2 then y must equal 18.
   Notice that a rather similar pair of equations is much easier to solve:

$$2x + y = 22$$
$$x/y = 2.25$$

Therefore

$$x/(22 - 2x) = 2.25$$

and so

$$x = 2.25 \times (22 - 2x)$$
$$x = 49.5 - 4.5x$$
$$5.5x = 49.5$$
$$x = 9$$

which means that $y = 4$.

Each of these pairs of equations could be solved graphically with similar ease (or difficulty). Plot $y = 22 - 2x$ against $y = 36 / x$ and $y = 22 - 2x$ against $y = x / 2.25$.

(Simultaneous equations can also be solved by matrix algebra, but we won't go into that.)

Solving equations by successive approximations

Consider this equation:

$$x / \log x = 14.11$$

The value of x to two places of decimals is wanted.

Writing this as $x = 14.11 \times \log x$ or as $\log x = x / 14.11$ doesn't lead any further towards a solution, i.e. values for x that satisfy the equation.

However, drawing a graph of the equation $y = x / \log x$ and reading off the values of x when $y = 14$ would give close approximations to the solution.

To draw such a graph you might start by noticing that when $x = 10$ then $y = 10$, when $x = 100$ then $y = 50$ and when $x = 1000$ then $y = 333.3$. Hence, to satisfy the equation it seems that x must be >10 and <100.

To examine the first possibility try a guess of $x = 30$; y then evaluates as 20.31, so that $x < 30$. When $x = 20$, y is 15.37, which is still too big. When $x = 15$, y is 12.75, which now is too small.

Thus, we know that

$$x \text{ is } >15 \text{ and } <20$$

More approximations will show that $x > 17$ and < 18. Still more approximations will give x to the required accuracy.

It's a slow process, but it does give the answer: $x = 17.56$.

Only if you have to make many calculations of this kind might it be worthwhile to spend time (probably a lot of time) in writing a computer program that can do the repeated iterations very quickly.

Unfortunately, the equation I have given to illustrate successive approximations is not an ideal example because the equation is also satisfied when $x = 1.22$. This would have been revealed if you had drawn the graph of $y = x / \log x$. The second solution can be reached, again by successive approximations, starting between the limits $x = 1$ and $x = 10$.

The value of y (i.e. $x / \log x$) has minimum at about +6.26 for all values of $x \geq 1$. If you can see what is the value of x at this minimum value of y then you will see why the minimum is *about* (hint) 6.26. (Another hint: $\log x = 0.4342 \ln x$.)

4.4 Rearranging equations

Several instances have already been given. Usually the object is to define an unknown (say a, which is part of an equation defining another unknown, say z) in terms of the other quantities in that equation.

For example, $z = 3a - b + 7$; define a in terms of b and z:

$$z + b - 7 = 3a$$
$$a = (z + b - 7)/3$$

and if you want b defined:

$$z + b = 3a + 7$$
$$b = 3a + 7 - z$$

To check the correctness of a rearrangement you may put arbitrary values into the starting equation, and see whether the rearrangement gives the right answers. Thus in the equation $z = 3a - b + 7$ let $a = 1$ and $b = 2$, which means that z must be 8. Now put z and b into the equation for a, and see if the right value for a comes out:

$$8 + 2 - 7 = 3; \quad 3 / 3 = 1 \text{ OK}$$

Similarly, $b = 3 + 7 - 8 = 2$ again OK.

A more tricky rearrangement is needed for an equation such as

$$z = \frac{ab}{2} + a$$

(if $a = 1$ and $b = 2$ then $z = 2$)

The danger here is in rewriting this as:

$$z = \frac{a(b + 1)}{2}$$

which is wrong (try it: z would evaluate as 1.5, not 2, as it should). The correct first move is to put

$$z = \frac{ab}{2} + \frac{2a}{2} \quad \text{so that } z = \frac{a(b+2)}{2}$$

Now, $2z = a(b + 2)$, and so $a = 2z / (b + 2)$ (Check $1 = 4 / (2 + 2) = 1$, OK)

Similarly $2z / a = b + 2$, and so $b = 2z / a - 2$ (Check $2 = 4 / 1 - 2 = 2$, OK)

Rearrangements are not always so easy – in fact they may be difficult! For instance, convert the equation $D = \mu_{max} [s / (K_s + s)]$ (from continuous culture) into a form that demonstrates that plotting $1 / D$ against $1 / s$ should give a straight line of intercept $1/\mu_{max}$ on the $1 / D$ axis and intercept $-1 / K_s$ on the $1 / s$ axis.

$$D = \frac{\mu_{max} \cdot s}{K_s + s} \quad \text{so } D(K_s + s) = \mu_{max} \cdot s$$

$$\text{and } (K_s + s) = \mu_{max} \cdot s \times 1 / D$$

All we have done so far is to take the reciprocal of the starting equation

$$\frac{1}{D} = \frac{K_s + s}{\mu_{max} \cdot s}$$

We now write this as

$$\frac{1}{D} = \frac{K_s + s}{\mu_{max}} \times \frac{1}{s}$$

which is the same as

$$\frac{1}{s} \times \left(\frac{K_s}{\mu_{max}} + \frac{s}{\mu_{max}} \right) \times \frac{1}{s}$$

Cancelling and rearranging gives

$$\frac{1}{D} = \frac{K_s}{\mu_{max}} \times \frac{1}{s} + \frac{1}{\mu_{max}}$$

When $1 / D$ represents y and $1 / s$ represents x you see that this is the equation of a straight-line graph with an intercept on the y ($1 / D$) axis of $1 / \mu_{max}$.

To find the intercept on the x axis we set $1 / D$ to zero, in which case:

$$\frac{K_s}{\mu_{max}} \times \frac{1}{s} = \frac{-1}{\mu_{max}}$$

Multiplying each side by μ_{max} leads to $K_s \times 1/s = -1$ and dividing by K_s finally gives

$$1/s = -1/K_s \text{ (only when } 1/D = 0)$$

Now try to rearrange the equation $D = \mu_{max} [s/(K_s + s)]$ into these three forms $\mu_{max} = ?$; $K_s = ?$; $s = ?$ so that μ_{max}, K_s or s respectively do not appear on the right-hand side of the equations. Not too hard, is it? See Chapter 9.

5 | Logarithms: exponential and logarithmic functions

My lord, I have undertaken this long journey purposely to see your person, and to know by what engine of wit or ingenuity you came first to think of this most excellent help in astronomy, viz. the logarithms, but, my lord, being by you so found out, I wonder nobody found it out before, when now known it is so easy.

Briggs to Napier

The aim of this chapter is to explain what logarithms are, and to discuss the ways in which they are used. Some of the material is not very easy to grasp, but you do need to have an understanding of this area. Even for people who already are familiar with logarithms there is probably something new in this chapter.

Logarithms

A logarithm is a way of writing one number (x) expressed as a power (index) of a second number (y) which is called the base, and which must be a real number >1. Some examples should make clear what this means. The number 8 is 2^3, and therefore if 2 is used as the base we can write: $\log_2 8 = 3$; in words this is to say that the logarithm of 8 to the base 2 is 3. Now, if 8 rather than 2 had been used as the base then $\log_8 8 = 1$ ($8 = 8^1$). If 64 were the base, then 8 ($=\sqrt{64}$) would be expressed as $\log_{64} 8 = 0.5$. If 3 were the base, then 9 (3^2) could be expressed as $\log_3 9 = 2$, and 27 (3^3) as $\log_3 27 = 3$. **Citing the logarithm of a number is only meaningful when the value of the base is also quoted or is clearly implied.**

5.1 Things to realise about logarithms

A logarithm is a pure number

The number that a logarithm represents might be (for example) moles per litre or cm per s, or almost anything else, but the logarithm itself has no units. What you can write is $\log_{10} (\text{cm s}^{-1}) = 2$, which would mean that $\text{cm s}^{-1} = 100$

Zero and all real numbers less than zero have no logarithm

The logarithm of 1 (to any base) is zero; $\log_{10} (0.1) = -1$; $\log_{10} (0.001) = -3$; $\log_{10} (1 \times 10^{-n}) = -n$. Hence, while a logarithm may have a negative value, **it always represents a number greater than 0**, no matter how negative the logarithm may be (nor what is the base of the logarithm). This means that on a graph with a logarithmic scale a zero or negative number cannot be plotted.

5.2   Bases used for logarithms

In principle, *any* positive real number greater than 1 could be used as the base to make a table of numbers accompanied by their logarithms. In practice, the only numbers that are used as bases for published tables are 10 (e.g. $\log_{10}$ $5 = 0.6991$; $\log_{10} 1000 = 3$) and 2.71828 … (this bizarre number is called e and there are good reasons, as we shall see later, for its use as a base). Logarithms (of a number, x) to base 10 are called **common logarithms** or just 'logarithms' and are written as $\log_{10} x$ or (more usually) just $\log x$, while logarithms to base e are called **natural logarithms** (or sometimes 'Napierian logarithms') and are written as $\log_e x$ or (more usually) $\ln x$.

Many people find it much easier to think about their money gaining interest in a bank than they do about numbers of microbes increasing in a batch culture. Therefore, we will have a financial digression.

If you have £100 invested in shares that give a 10% dividend annually, then every year you will receive £10 interest. This is called **simple interest**.

If you have £100 in a deposit account that adds 10% interest every year to the sum deposited, after one year you will have £110. However, next year (if you keep all the money in the bank) you will have 10% of £110 as interest, which is £11, so that there is now £121 on deposit. Next year the interest will be £12.1, and you will have £133.1 on deposit. Each year the amount of interest and the total value of the deposit grow. This is called **compound interest**. The rate of growth of your money (the interest gained in any one year) is directly proportional to the total of the amount of your initial deposit plus the already accumulated interest.

Notice carefully: (1) Saying that a positive real number (n) is increased by 10% is the same as saying $n \times 1.1$. Likewise, a 5% increase $= n \times 1.05$, and so on. (2) How much n increases when raised by 10% depends on the actual size of n. A 10% increase of £10 is only £1, while a 10% increase of £10 000 is £1000. If you get 10% compound interest annually for x years then your initial deposit (n) will become $n \times 1.1^x$. So, after 10 years you would have $n \times 1.1^{10} = 2.593\ n$.

All this is familiar (I hope) but these concepts need to be very clearly understood.

Now things get a bit harder to follow. What if instead of 10% once annually you were given compound interest of 10 / 12% monthly? Again note carefully that 10 / 12% monthly (an increase by a factor of 1.008 33 each month) is called an equivalent rate to 10% annually, but the outcome is not the same because $n \times 1.008\ 33^{12} = 1.1047\ n$. If you got this monthly rate for 10 years then you would finish with $n \times 1.008\ 33^{120}$ which is $2.707\ n$. Again note that the amount of money you actually get depends on the size of n.

Another equivalent (to 10% annually) rate is 10 / 365% given daily. This yields after 10 years $n \times 1.000\ 273^{3650} = 2.718\ n$. We could go on calculating interest given hourly, and then at every minute at rates 'equivalent' to 10% annually. What we would be doing in effect is evaluating $(1 + 1/x)^x$ as x becomes a larger and larger number. The great mathematician Euler proved that as x increases this expression leads towards the irrational number 2.718 28 … which is called e. If interest were given *continuously* at a rate equivalent to 10% annually for 10 years then you would end with $n \times e$.

No bank does give interest continuously (though they do charge interest monthly for debts on credit cards!), but growth of a microbial batch culture does come very close to being continuous. If you have 1 000 000 organisms at various stages of their division cycle, and a doubling time of one hour (3600 s) then in every second roughly 1 000 000/3600 $= 278$ organisms will divide.

5.3   How Napier discovered logarithms, and why logarithms to base e are natural

With the idea of finding how money accumulated as it earns compound interest, John Napier (1550–1617) multiplied 1 by 1.0001 (equivalent to

Table 5.1 Examples of the successive results of repeatedly multiplying by 1.0001, starting with 1 × 1.0001 (step 1 in the table), then 1.0001 × 1.0001 (step 2) etc.

Step	Result
1	1.0001
2	1.0002
3	1.0003
5	1.0005
10	1.0010
100	1.0101
500	1.0513
1 000	1.1052
2 000	1.2214
3 000	1.3499
5 000	1.6487
10 000	2.7183

0.01% interest). He wrote the answer to this step 1 (= 1.0001) and then multiplied the result by 1.0001. This gave the answer to step 2, which he recorded. The step 2 answer in turn was multiplied by 1.0001 to give the step 3 answer. Napier went on in this way for many successive steps, each time multiplying the answer by 1.0001 to produce a table of the results (illustrated in Table 5.1). (This isn't quite what he really did, but my story is easier to understand, I hope.)

The remarkable thing that Napier recognised was that if he took two of his step numbers, for example 1000 and 2000, added them together (to give 3000 in this instance) and looked in his table for the value corresponding to this sum of steps he found that the value (1.3499 here) was equal to (1.1052, the step 1000 number) × (1.2214, the step 2000 number) = 1.3499. This way of multiplying two numbers worked no matter which two step numbers were added. Napier also found that subtracting step numbers was equivalent to division. Thus, step 5000 minus step 3000 gives step 2000, which corresponds to 1.2213, which is the answer to 1.6487 (step 5000) / 1.3499 (step 3000).

Napier coined the word logarithm to describe his step numbers, from two Greek words, *logos* (reckoning) and *arithmos* (number). He first applied his new idea as a primitive slide rule, which was called 'Napier's bones'. It was obvious too that a table of Napier's step numbers with their corresponding values gave an easy way of turning multiplications and divisions into additions and subtractions. Napier had no conception of e ('the base of natural logarithms' $= 2.718\,18\ldots$), nor of the idea of a base at all, but he had found a wonderful way of simplifying arithmetic. Because they were not determined from any preconceived base, his logarithms were 'natural'. We can see that Napier's steps correspond (very nearly) to powers of e: step $1 = e^{0.0001}$; step $100 = e^{0.01}$; step $1000 = e^{0.1}$; step $10\,000 = e^{1}$ and so on. Because the steps are indices then adding steps is equivalent to multiplication.

For nearly 400 years people have used logarithms, mostly with no understanding of how they were derived or how they work. However, over this long period of time mathematicians have increased our knowledge in this area, and have produced the many rules and relationships that we are considering in this chapter.

5.4   Calculation of natural logarithms and their conversion to logarithms of other bases

You will probably never in your life need to calculate the logarithm of a number. Pocket calculators and computers are preprogrammed to make the calculation at the touch of a button. The following paragraph tells you how it is done.

To find the natural logarithm of n (any positive real number) one may use the series:

$$\ln n = 2 \times (a^{1}/1 + a^{3}/3 + a^{5}/5 + a^{7}/7 + a^{9}/9\ldots)$$

where $a = (n - 1)/(n + 1)$. The terms in a series may continue indefinitely, but may converge towards a limit if they become progressively smaller, as is the case here. How many terms need to be evaluated to give a satisfactory sum for the series will depend on the precision required in the answer, and on the value of n: small values give a quicker answer. (Do you see a way of starting the series with a small value of a (hint: $a < 1.718 / 3.718$) no matter what the value of n?)

It is simple to convert the natural logarithm of a number into the logarithm of that same number to any other desired base (see later).

5.5   Antilogarithms and their calculation

We have seen how the natural logarithm of a number may be calculated, that is, given a positive real number n, we can find the value of x that makes $e^x = n$. Very often we need to do the converse of this operation, that is, given a value x (positive or negative) which is a natural logarithm (i.e. power of e) we want to find the number n for which $n = e^x$. This process is called finding the antilogarithm, and n is said to be the antilog (to the base being used) of x. Once again, we would use a pocket calculator to do the work for us, but the following paragraph shows how it is done.

To convert e^x into n, we use the series:

$$e^x \; (= n) = 1 + x + x^2/2! + x^3/3! + x^4/4! + x^5/5! + x^6/6! + \dots$$

Here 2!, 3!, 4!, 5! etc. mean factorial 2 (1×2), factorial 3 ($1 \times 2 \times 3$), factorial 4 ($1 \times 2 \times 3 \times 4$), factorial 5 ($1 \times 2 \times 3 \times 4 \times 5$), factorial 6 ($1 \times 2 \times 3 \times 4 \times 5 \times 6$) etc. To evaluate e^{-x} we determine e^x, and then find $1 / e^x$.

5.6   Conversions

If we know the value of the logarithm of a number x to one base (b) then it is possible to find the logarithm of x to any other desired base (c) from the relation:

$$\log_c x = \log_b x / \log_b c$$

The particular usefulness of this equation is that it allows us to convert natural logarithms (which are relatively easy to calculate) into logarithms of any other required base. With the equation we can also convert common logarithms (which are easy to tabulate) into logarithms of a different base. Some examples follow:

$$\log_{10} n = \ln n / \ln 10 = \ln n / 2.302585 \dots \text{or} = 0.43429 \; \ln n$$

(hence $2.302585 \times \log_{10} n = \ln n$ which is a frequently used relationship).

$$\log_2 n = \ln n / \ln 2 = \ln \; n / 0.693\,147 \dots$$

$$\text{or,} \quad \log_2 n = \log_{10} n / \log_{10} 2 = \log_{10} n / 0.3010$$

$$\text{e.g. } \log_3 8 = \ln 8 / \ln 3 = 2.079\,442 / 1.098\,612 = 1.89279$$

To find the antilog when e was not the base of our logarithm, we can first convert to a natural logarithm by multiplying with an appropriate factor. In practice, if we are not using natural logarithms, we shall almost certainly have logarithms to base 10, and in that case we take the relation:

$$\ln n = 2.302\,585 \times \log_{10} n$$

to calculate $\ln n$ and hence determine the antilog.

Determining the value of e

Setting x to 1 in the equation for antilogs gives the value of e^1 which is e itself. With only 10 terms evaluated and summed we find that $e = 2.718\,282$ and the 10th term is only 0.000\,003. Higher terms quickly become smaller and smaller, which means that *to five places of decimals e* does not get any bigger no matter how many more terms are added. However, we could go on and on putting figures to the extreme right of the decimal by computing and adding higher terms. It should be obvious, therefore, why e is an irrational number – we can never precisely express its value. To 15 decimal places it is 2.718\,281\,828\,459\,046, but figures further to the right can be added, without recurring, *for ever!* The immediate repetition of the digits 1828 is remarkable because the sequence of digits of e is entirely random.

From what has been said so far, it may seem that this weird number e is used only because logarithms and antilogarithms to this special base are easy to calculate. There are other, and more important reasons for using e and natural logarithms, as will be seen later.

Logarithms to base 10

Our counting system is based on 10, and for this reason common logarithms have a special place. Any positive real number can be written as $n \times 10^x$, e.g. $3790 = 3.79 \times 10^3$; $0.0543 = 5.43 \times 10^{-2}$and it is easy to make such conversions by inspection. In consequence, common logarithms needed to be tabulated for only the numbers 1 to 10. The logarithm of any number then can be found from the table by finding $\log_{10}$ of a number from 1 to 10 and then adding to it the correct power of 10 (of our number). Thus,

$$\begin{aligned}
\log_{10} 54\,073 &= \log_{10}(5.4073 \times 10^4) \\
&= \log_{10} 5.4073 + \log_{10} 10^4 \\
&= 0.732\,98 + 4 = 4.732\,98
\end{aligned}$$

And $\log_{10} 0.05 = \log_{10}(5 \times 10^{-2})$
$= \log_{10} 5 + -2 = 0.699 - 2 = -1.301$

Now that calculators have supplanted tables of logarithms, there are fewer occasions when it is really necessary to use logarithms to any base other than e. We may often choose still to use common logarithms because it is so easy to tell (just by looking) roughly the size of the number that the logarithm represents. Thus, if 3.4376 is the $\log_{10}$ of a number, we know at once that the number itself is >1000 and <5000 (0.4376 is less than 0.699 which is the $\log_{10}$ of 5). On the other hand it is not so easy to tell at a glance what is the rough size of the number for which the ln is 11.5405: in this case the number is approximately 103 000. Would you have been able to make a close guess?

Logarithms to base 10 are helpful only because of our number system. You should realise that the **value of e is independent of a number system; e** is 2.718 28 ... in binary, hexadecimal or any other system, in the same way that π is always 3.141 59 ... In going from one number system to another, the way that e is written will change, but the numerical value of e is unchanged.

5.7   Some uses of logarithms

Before the appearance of cheap calculators, common logarithms were very much used to make multiplications and divisions easier to do. This was because:

The logarithm (to a given base) of the product of two (or more) numbers is equal to the sum of the logarithms (to the same given base) of the numbers.

$$\log_c(a \times b) = \log_c a + \log_c b$$

The logarithm of the quotient of two numbers is equal to the difference of the logarithms of the numbers.

$$\log_c(a/b) = \log_c a - \log_c b$$

(Both of these equations follow from the rules of indices, and logarithms are indices.)

Everyone now has a pocket calculator that will do multiplications or divisions and there is no need to use logarithms for these operations. Nevertheless, logarithms retain their value in many other calculations.

The logarithm of a number raised to a power is equal to the logarithm of the number multiplied by the power.

$$\log_c(a^b) = b \times \log_c a$$

This last relation is very important since it affords an easy way to evaluate expressions like $7^{4/3}$ or $2^{3.507}$, or $5^{-1/4}$ thus:

$$\ln 7 = 1.945\,910$$

$$\therefore \ln 7^{4/3} = (4 \times 1.945\,910)/3 = 2.594\,546$$

$$e^{2.594546} = 13.3905 = 7^{4/3}$$

$$\ln 2 = 0.693\,15$$

$$\therefore \ln 2^{3.507} = 3.507 \times 0.693\,15 = 2.430\,87$$

$$e^{2.430870} = 11.3687 = 2^{3.507}$$

$$5^{-1/4} = 1/5^{1/4}.\ \ln 5 = 1.609\,44.$$

$$1.609\,44 \times 1/4 = 0.402\,36.$$

$$e^{0.402360} = 1.495\,35$$

$$\therefore 5^{-1/4} = 1/1.495\,35 = 0.668\,74$$

Many calculators have a button labelled 'x^y' which allows you to solve the above kinds of expression very easily. You enter x, press x^y, then enter y (made negative if necessary), and the calculator produces the answer, having used the method just described.

If your calculator can solve x^y as well as doing multiplications and divisions, you might think that you have no reason to concern yourself with logarithms. As a scientist you would then be quite wrong.

Exponential growth of microorganisms

When bacteria are put into a flask of fresh medium there is a lag phase and then a period of unrestricted growth. During this unrestricted period the rate at which the number of organisms increases (expressed as, for example, new organisms produced after incubation for 1 hour) depends on how many bacteria were present at the start of the 1-hour period. Thus, if we started with 100 organisms we might perhaps find 180 after 1 hour, whereas if we had started with 5000 of the same organisms then we would expect many more than 80 new bacteria to be formed in 1 hour; instead we might now find 4000 new organisms (that is, 9000 in total minus the 5000 present initially). In a further 1 hour these 9000 might become about 16 000, which is an increase of 7000 in 1 hour.

We can say that, during unrestricted growth, the rate of increase at any moment will be proportional to the number of organisms present at that moment. Mathematically this state of affairs is expressed as:

$$dn/dt = kn$$

In general, dy/dx means the minute increase of a quantity y for an infinitesimal increase of a related quantity x, that is the rate of increase of y relative to x. Here dn/dt means the increase of n (i.e. dn) during an exceedingly short time (that is dt). Hence, the equation is stating that the rate of increase of n at any given moment is equal to the number of organisms then present (= n) multiplied by a constant, k. This is what **exponential growth** means – it does not *necessarily* mean fast growth. The constant k is the same thing as percentage increase of n at the same given moment.

Knowing the gradient of a graph (dn/dt) does not tell us what is the equation that has led to the graph. We need such an equation so that we can calculate n_t, the number of organisms after an elapsed time t, when there were n_0 organisms at time 0.

To find the equation start by taking k to be 1.

In such circumstances, we get:

$$dn/dt = n$$

In other words, the rate of change of n (number of organisms) with respect to t (time) might be numerically equal to n itself. Should this appear a strange idea, then realise that if you were so fortunate as to be given 100% interest (once, at the end of a year) then after one year the increase in your money would equal the amount you originally had deposited.

It follows from the equation $dn/dt = n$ that a graph of n against t must have a slope which at any point is equal to the value of n at that point. Is there an equation of the kind $y = f(x)$ (that is to say can y be defined as a function of x; i.e. is there an equation to allow us to calculate y if we know x, or if we know y then to calculate x) that will lead to a graph on which **the slope (dy/dx) at *every* point on the curve is equal to the value of y at that same point?**

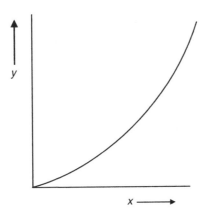

Fig. 5.1 The shape that a graph must have if y is to be equal to dy/dx at all values of y.

Obviously any equation that gives a straight-line graph cannot be our required $f(x)$ because the slope of all such graphs is fixed and does not alter as y changes. Likewise, any equations that yield curves with inflexions (maxima or minima) cannot meet the requirement because there must then inevitably be occasions when y is decreasing as x increases. What is needed is an equation that plots as a curve like the one shown in Fig. 5.1.

A shape of this kind is given by equations such as $y = x^2$ or $y = x^3$ (in general $y = x^a$ and $x > 1$), but the slope of any such graph is not equal to y *all the time*. The slope of all these graphs is given at any point by $a.x^{(a-1)}$ and this will almost never equal y.

Another kind of equation that will produce a graph of the desired shape is $y = c^x$, where c is defined as a positive (but not necessarily integral) constant real number greater than 1. In Fig. 5.2 graphs are shown of $y = c^x$ for values of c of 1, 2, 3 and 4, and values of x from 0 to 1. When $c = 1$ then y has an unvarying value of 1 no matter what is the value of x, and dy/dx is always 0. When $c = 2$ then y increases as the value of x rises, but at any point y is always bigger than dy/dx. However, when $c = 3$ then y becomes always smaller than dy/dx, and when $c > 3$ then y becomes considerably smaller than dy/dx. Hence, it looks as though there may indeed be a value for c, closer to 3 than to 2, for which y and dy/dx are equal for all values of y. By now it should be easy to guess that this value of c is going to be e, that is 2.718 28 ...

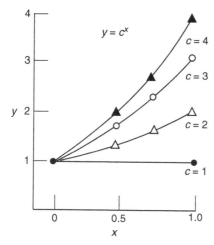

Fig. 5.2 Graphs of $y = c^x$ for some integer values of c.

As a result of all this, we find that the equation:

$$y = 2.718\,28^x \ (= e^x)$$

does **uniquely** (mathematicians have proved) satisfy the requirement that the rate of increase of y (dy/dx) is equal to y. This equation can be regarded as a special case of a more general relation:

$$y = a.e^{b.x}$$

where a and b are constants (both equal to 1 in the equation $y = e^x$). This is the form of equation that we need to relate n (number of bacteria) to time of growth.

Again, some examples may make this easier to follow.

How much would £100 be worth after 2 years if interest were given continuously (not monthly or even daily) at a rate equivalent to 5% per year? Here $a = 100$ and $b = 0.05$ while $x = 2$. So, $y = 100 \times e^{0.05 \times 2} = £110.52$.

How many organisms will be present after exponential growth for 3 hours of an initial population of 5000 bacteria, with a specific growth rate of 0.011 per minute? Here $a = 5000$ and $b = 0.011$ per minute, while $x = 180$ (b and x must be expressed in the same units). Hence y (organisms after 3 hours) $= 5000 \times e^{0.011 \times 180} = 36\,214$.

The number of organisms (n_t) present after exponential growth for time t ($\equiv x$) is:

$$n_t = n_0 \times e^{\mu t}$$

where n_0 ($\equiv a$) is the number present initially, and μ ($\equiv b$) is the specific growth rate.

The doubling time (t_d) is the time needed during the exponential phase for n_t to become $2 \times n_0$, and so (in this special case):

$$n_t/n_0 = 2 = e^{\mu t_d},$$
and consequently $\mu.t_d = \ln 2 = 0.6931$

This means that $\mu = 0.6931 / t_d$ or that $t_d = 0.6931 / \mu$.

We can determine μ if we know t_d, or t_d if we know μ, but where should we start? Either μ or t_d must be found by experiment, as discussed below.

From the equation $n_t = n_0 \times e^{\mu t}$, it follows that $\ln n_t = \ln n_0 + \mu t$. This can be rearranged to $\ln n_t = \mu t + \ln n_0$, and this shows that a graph of $\ln n_t$ (y axis) against t (x axis) will yield a straight line during the exponential phase of growth, of slope μ and intercept $\ln n_0$ on the y axis. Most frequently, one would determine the number of organisms (per unit volume) or some other property that is directly proportional to number (such as optical density) and plot the $\ln$ of this against the time at which each measurement was made. The slope of the straight line part of the graph ($\Delta \ln n / \Delta t$) will be equal to μ. Logarithms are pure numbers (not numbers *of* anything) and so the units of μ are a pure number per unit time (e.g. 0.015 min^{-1} or 0.90 h^{-1}).

We could find μ by algebra, without a graph, if we know the time interval (t) between which two measurements had been made of numbers of organisms (n_1 and n_2) per unit volume:

$$n_2 = n_1 e^{\mu t}$$
$$\therefore \ln n_2 = \ln n_1 + \mu t$$
$$\therefore (\ln n_2 - \ln n_1)/t = \mu$$

The drawback to this procedure is that it presupposes that both n_1 and n_2 were indeed measured during the exponential phase of growth. The only simple way of establishing whether *growth* really is exponential is to plot log $_{(any\ base)}$ n against time and see whether the points lie on a straight line (of positive gradient). We must have at least three values of n to do this (since a straight line can always be drawn through two points).

Decay of radioisotopes

The rate of decay of a radioisotope cannot be altered by chemical or physical means, and is dependent only on the composition of the nucleus. The number of nuclei disintegrating in unit time will also be proportional to the number of radioactive nuclei initially present. Mathematically, the rate of decay is given by:

$$- dn/dt = kn$$

where n is the number of radioactive nuclei, t is the time and k is the **decay constant**, characteristic of the particular radioisotope that is being used. This equation is clearly very similar to the one that represents exponential growth of microorganisms; the difference is that now the number of atoms decaying has to be *subtracted* from the number of atoms initially present (whereas the number of new bacteria growing was added to those initially present) and it leads to the next equation:

$$n_t = n_0 \, e^{-kt}$$

by the same kind of reasoning as has already been discussed. Here n_0 and n_t are the numbers of radioactive nuclei at times 0 and t respectively.

The **half-life** (t_h) of the radioisotope is the time needed for n_0 to fall to half of its initial value, and so:

$$n_t/n_0 = 1/2 = e^{-kt_h} = 1/e^{k.t_h}$$
$$\therefore e^{k.t_h} = 2$$
$$\therefore kt_h = \ln 2 = 0.6931, \quad \text{and}$$
$$t_h = 0.6931/k \text{ or } k = 0.6931/t_h$$

The half-life can be found graphically by plotting the ln of the number of nuclei disintegrating ($\equiv$ dpm) against time. This should give a straight line (Fig. 5.3) from which the half-life can be read by finding the time interval during which ln dpm decreases by 0.693 (= ln 2). The value of k, the decay constant, can then be deduced, or it can be found directly from the gradient of the line, which is equal to $-k$.

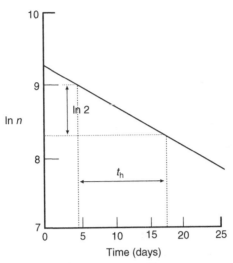

Fig. 5.3 Radioactive decay of ^{32}P. The radioactivity (n dpm) of $Na_2H^{32}PO_4$ was measured at daily intervals. The half-life (14 days) is found from the time during which ln n decreases by ln 2. The decay constant is therefore $0.6931/14 = 0.0495$ day^{-1}.

The average lifetime of a radioactive atom can be shown mathematically to be the reciprocal of k, and so for ^{32}P is 20.2 days. Because k, the half-life and the average life are so simply related it is sufficient to quote any one of these numbers to define the stability of a radioisotope. The half-life is now the value most often given, even though k is more directly useful in calculations, and the average life is almost never mentioned.

5.8   The logarithmic function

We have seen that the exponential function is:

$$y = e^x$$

(which is the same as $y = $ antilog$_e$ x). The logarithmic function is the reverse of this, namely;

$$y = \log_e x, \text{ or more generally}$$
$$y = \ln(ax^b), \text{ which is to say that } y = b.\ln x + \ln a$$

A graph of y against ln x will give a straight line of slope b and intercept ln a on the y axis.

The exponential function is recognised as being important by micro-biologists. The logarithmic function is also important in biology, but its applications may go almost unnoticed. It is often used to describe the relation between the responses of organisms, y (= growth or inhibition for instance), to a wide range of concentrations (x) of an administered substance, usually expressed as $\log_{10} x$. In particular, and very importantly, **we use the logarithmic function in defining the pH scale of hydrogen-ion concentrations**. The pH value of a solution is the negative of the logarithm (always to base 10) of the molar concentration of hydrogen ions.

$$\text{pH value} = -\log_{10}[H^+\text{concentration (moles } L^{-1})]$$

For instance, suppose $[H^+] = 2$ mM. This concentration is 2×10^{-3} M (0.002 M), and so $\log_{10} [H^+] = -2.6990$, and therefore the pH value is 2.699.

Caution! Be very careful about taking the average of numbers that are from a logarithmic scale. For example, suppose that you determine the pH values of two cultures (of the same organism in the same medium) when growth has reached the stationary phase, and get values of 8.0 and 8.8. The average is apparently 8.4, but this is incorrect. A pH value of 8.0 represents 1×10^{-8} M H^+ and a value of 8.8 represents 0.16×10^{-8} M H^+ so that the average is 0.58×10^{-8} M H^+ which means the correct average pH value will be 8.2. This is just like showing that the average of 1×10^1 (10), 1×10^2 (100) and 1×10^3 (1000) is not 1×10^2 but is really 3.7×10^2.

Another application of the logarithmic function is in defining the extinction (E) (or absorbency) of a solution that absorbs light:

$$E = \log_{10} (I_0/I)$$

where I_0 and I are the intensities of the incident and the transmitted light respectively (at a specified wavelength). To show its relation to the general form of the logarithmic function the equation can be written as $E = \ln (I_0/I) \times 0.4343$ ($E \equiv y$; $I_0 \equiv a$; $I \equiv x$; $-1 \equiv b$; the factor 0.4343 converts the natural logarithm to $\log_{10}$). Because E (though a logarithm) is directly proportional to the concentration of a coloured solute you can validly determine the average of several values of E. For example $E = 0.4$ might represent 0.1 μmol ml^{-1} and $E = 0.8$ will then represent 0.2 μmol ml^{-1} (see Chapter 9 about spectrophotometry). The average E of 0.6 will represent 0.15 μmol ml^{-1}, which is correct.

Many biologists use the words 'exponential' and 'logarithmic' (when applied to a certain phase of growth of microorganisms in batch cultures) as though they had the same meaning and were interchangeable. This is wrong, because the words really have quite different meanings, and one ought always to speak of the 'exponential phase' of growth. Also, when this wrong term 'logarithmic phase' is abbreviated in speech to 'log phase' there is a possibility of mishearing these words as 'lag phase'.

No one ever seems to make the mistake of describing a logarithmic function (such as the extinction of a coloured solution or pH value) as being an exponential function.

Semi-logarithmic graph paper

Do not use it! This kind of paper was introduced when looking up logarithms in tables was laborious. The paper allows numbers to be plotted in the positions of their logarithms (to base 10) on the graph without you needing to find the logarithms. Now that it is easy to get logarithms from a calculator most of the advantage has disappeared. The drawbacks remain.

Finding the gradient, in the correct units, is not entirely simple, even with a straight line.

Very often one does not have the best number of cycles on the paper, so that sheets have to be joined (if there are too few cycles) or a small graph has to be drawn (if there are too many cycles).

Many people do not understand what they are doing with this kind of paper. Mistakes in labelling the axes, and in plotting logarithms rather than numbers, are very common.

Semi-logarithmic paper costs more than ordinary graph paper!

Getting logarithms from your calculator and plotting them, as logarithms, on ordinary graph paper is a better practice, and it is easier to understand what you are doing.

An afterthought

An afterthought that seems to have no relevance at all to biology: one cannot leave the subject of logarithms without mentioning the extraordinary relation proved by Euler:

$e^{i\pi} = -1$ (where i is the square root of -1, and π is $3.142\ldots$)

The equation looks like a miracle of Nature (two irrational numbers together with i resolving to an integer), but the proof is fairly simple, because it is in fact no more than a special case of the general (but almost equally surprising) relation of e to the trigonometric functions: $e^{i\theta} = $ cosine $\theta + i$ sine θ (where θ is an angle measured in radians rather than degrees).

6 | Simple statistics

If an experiment in microbiology has been well designed, no statistics are needed.

D. D. Woods

You don't have to agree with all of my chosen quotations, and in this instance I disagree with my former supervisor. At the least, statistical treatment of data can help to convince yourself, and perhaps others too, of the reliability of your conclusions. Statistics are also important in cricket and bridge.

Often one repeats an estimation (notably of an enzymic activity) several times, and then reports the average result (the mean) and its standard deviation (that is, an indicator of how much the results scatter about the average). This is a **descriptive use** of statistics. A **predictive use** occurs when assessing the probability that two means (such as the activities of an enzyme in organisms grown under two conditions) are really different. In this chapter we shall see how to work out a standard deviation, and how to compare means. Going beyond these procedures (which are not all that simple) soon leads to a call for the services of a professional statistician, and damn the expense! It's like going to law.

Even determining the mean and its standard deviation is not straightforward because we have to take into account whether a mean is calculated from a whole population or from a sample of a population, and also whether the total number of individual values that contribute to a mean are small (in practice less than 30) or large ($\geq$30). There are other complications (discussed later) and the treatment in this chapter is very abbreviated, with little attempt to explain the mathematical theory which underlies the methods. A detailed and very lucid exposition is given in an excellent book: *Statistics: A First Course,* by J. E. Freund and G. A. Simon (see Further reading).

6.1 Simple statistical measurements

To begin, here are some definitions:

(a) **Average** ($\equiv$ **arithmetic mean**) Suppose we have n numbers, thus: $x_1, x_2, x_3, \ldots, x_n$. The mean of these numbers is their average:

$$(x_1 + x_2, +x_3, \ldots + x_n)/n$$

For example , consider these nine numbers: 5, 4, 5, 7, 6, 8, 9, 4, 7. Their sum is 55. Hence their mean is 55 / 9 = 6.11.

The mean is denoted by the symbol $\bar{x}$.

(b) **Median** The median is the middle number in a set of numbers, in the sense that half the numbers are less than the median, and half are greater. If we take the nine numbers from above and arrange them in ascending order: 4, 4, 5, 5, **6**, 7, 7, 8, 9 we find that 6 is the median. (When the total of numbers in the set is even, then the median is the average of the two middle numbers.)

(c) **Standard deviation** The above set of nine numbers has a mean of 6.11 and a median of 6. So too does this set: 5, 5, 5, 6, 6, 6, 7, 7, 8. However, the numbers in the second set are obviously more closely grouped about the mean. The **standard deviation** is a way of indicating the spread of a set of numbers about the mean, in the form of a single number, without having to show all the numbers (nine in this case) that contribute to the mean. Finding the standard deviation is nowadays very easy with a pocket calculator or with a spreadsheet program on a computer. However, you should understand how the standard deviation is evaluated.

6.2 Calculating standard deviation

First, we find out how much each individual value differs from the mean, i.e. $x_1 - \bar{x}, x_2 - \bar{x}$, etc. These differences are then each squared and the squared values (which all will be positive) are added together. Finally, this sum of the squares is divided by the total number of individual values that contributed to the mean. The result of this calculation is called the **variance**.

The variance is therefore:

$$[(x_1 - \bar{x})^2 + (x_2 - \bar{x})^2 + (x_3 - \bar{x})^2 + \ldots (x_n - \bar{x})^2]/n$$

It can be shown (not by me) that this is equivalent to:

$$[(x_1^2 + x_2^2 + x_3^2 + \ldots x_n^2)/n] - \bar{x}^2$$

This latter equation is easier to use, and it is a way that a pocket calculator can determine variance without having to store numerous separate value of x.

The standard deviation is the square root of the variance.

For the first set of numbers from above the variance is:

$$[(4^2 + 4^2 + 5^2 + 5^2 + 6^2 + 7^2 + 7^2 + 8^2 + 9^2)/9] - 6.11^2$$
$$= [(16 + 16 + 25 + 25 + 36 + 49 + 49 + 64 + 81)/9)] - 37.35$$
$$= (361/9) - 37.35$$
$$= 2.76$$

and the standard deviation is therefore $\sqrt{2.76} = 1.66$.

In the same way we can find for the second set of nine numbers that the variance is:

$$[(5^2 + 5^2 + 5^2 + 6^2 + 6^2 + 6^2 + 7^2 + 7^2 + 8^2)/9] - 6.11^2 = 0.983$$

and the standard deviation is therefore 0.991.

Notice particularly that the numerical value of the standard deviation alone is not what matters. The important point is the size of the deviation in relation to the mean. Thus, a deviation of 1.66 is 27% of a mean of 6.11 (while 0.991 is only 16% of the same mean value), whereas if the mean (of another set of numbers) had been 106 then a standard deviation of 1.66 would be only 1.6% of the mean.

In these calculations of variances (and hence of standard deviations) we used as a divisor n, the total (nine in these instances) of the numbers in the set. This is correct only when we are establishing the **variance of a whole population**, but not when finding the **variance of a sample** from a larger population. If our nine numbers were for example the ages in years of *all* the children in a very small school then the use of n would be right, but if the nine numbers had been the ages of *a sample* of nine children from a larger school then it would be wrong to use n. In this latter case, where we examine a sample and not the whole population, the correct procedure is to calculate the variance by using $n - 1$ as divisor rather than n. Hence we have the equation:

$$\text{sample variance} = [(x_1 - \bar{x})^2 + (x_2 - \bar{x})^2 + (x_3 - \bar{x})^2$$
$$+ \ldots (x_n - \bar{x})^2]/(n - 1)$$

The sample $(n-1)$ variance (and hence sample standard deviation too) will therefore be greater than the population (n) variance and population standard deviation. To calculate sample variance it is easiest first to calculate the variance of the numbers in our sample exactly as we did above, using n, as though we were dealing with a population, and then convert with this relationship:

$$\text{sample variance} = \text{population variance} \times n/(n - 1)$$

The sample standard deviation is of course the square root of the sample variance. As n becomes greater then the difference between the population and sample variances will become less.

A calculator or spreadsheet will give you the choice of determining variance and standard deviation with n (population) or $n-1$ (sample) weighting.

Work out the sample variances and standard deviations of the two sets of nine numbers from above. Now learn how to do this with a calculator. If you have access to a spreadsheet, such as Microsoft Excel® , then learn to use this as well.

In biology we seldom deal with all the individual numbers that make up a whole population. As instances, we may estimate the viable organisms in a culture by finding the average number of colonies that grow from perhaps three separate very small samples taken out of the culture; we may make five measurements of the activity of an enzyme in an extract and determine the average. In each case we have examined only a fraction of the huge number of estimates that could be made if life were not so short. Consequently, an enzymic activity will often be quoted in the form 30 [SD 4 $(n=3)$] nmol min^{-1} (mg protein)$^{-1}$. This means that the average of three estimates was 30 nmol min^{-1} (mg protein)$^{-1}$, with a sample standard deviation of 4 nmol min^{-1} (mg protein)$^{-1}$.

Suppose a mean of samples has been determined as above with $n=3$. It would not be at all surprising if three more samples (from the same population, assayed by the same method) gave a somewhat different mean, and a further batch of samples might produce yet another mean. On the other hand if a mean had been determined with $n = 50$ we would not expect to find a very different mean for a further 50 samples. How sure can we be that a sample mean is a good measure of the population mean? The answer is that we can calculate, from one mean and its standard deviation together with the number of values that contributed to the mean, the limits between which the true mean will lie 19 times in 20 (95% probability), or, if we want, 99 times in 100 (99% probability). In finding the confidence limits, as described below, some assumptions are made about the way individual results are distributed about the mean. These assumptions will be considered later.

6.3   Confidence limits

Rather than quoting the sample standard deviation of a mean, it used to be common to give the confidence limits of the mean. For example, the enzymic activity given above might be presented in the form 30 ± 11 nmoles min^{-1} (mg protein)$^{-1}$. This says that the mean found was 30 and that there is a high probability (likely to be 95% if not stated explicitly) that the true population mean (i.e. the average mean of very many estimates) does lie between 19 and 41 nmoles min^{-1} (mg protein)$^{-1}$. (It also implies a 5% probability that the true mean is outside these limits.) Such **confidence limits** are determined when $n \geq 30$ from the equations:

$$\bar{x} - (z \times \text{sample standard deviation})/\sqrt{n} \text{ and}$$
$$\bar{x} + (z \times \text{sample standard deviation})/\sqrt{n}$$

where z is a number, found from statistical tables, that establishes the level of probability for the limits: for 95% probability z is 1.96; for 97.5% probability z is 2.24 and for 99% probability z is 2.58. Determining 95% confidence limits is a two-tailed test (see later) and so z has the value 2.24 (2.5% applied twice).

When $n < 30$ a different equation must be used to find the limits:

$$\bar{x} - (t \times \text{sample standard deviation})/\sqrt{n} \text{ and}$$
$$\bar{x} + (t \times \text{sample standard deviation})/\sqrt{n}$$

Here the value of t (for a given probability) is not constant, but depends on the value of $n - 1$, and so must be looked up in statistical tables. The smaller is n, the larger is t; thus for $n = 3$, t (95% probability) = 4.30: t (97.5%

probability) = 6.21; and t is 9.92 for 99% probability. When $n = 10$ then t (95% probability) is 2.23 and is 3.17 (for 99% probability).

In quoting confidence limits their probability should also be given, but often are not, and 95% must then be assumed. The confidence limits, given alone, do not reveal the value of n, from which the mean and its limits were found. For these reasons showing the mean and standard deviation, together with the value of n, is preferable. Confidence limits then can be calculated if they are wanted. Notice particularly that in working out confidence limits at 95% probability we must use the z or t value that gives 97.5% probability. This is because there is then only a 2.5% probability of the true mean being small and similarly only a 2.5% probability that the true mean is large – these two probabilities together being only 5%. This is a **two-tailed test** (see later).

6.4   Normal distribution

A limitation of these very standard methods is that they require the graph of the distribution of results about the mean to be, at least approximately, a bell-shaped curve (Fig. 6.1).

While this distribution is very commonly found, it is easy to think of cases where the distribution is quite different, e.g. the scores of a good batsman

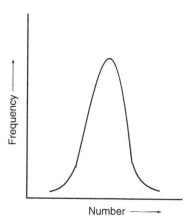

Fig. 6.1 The normal, or Gaussian, distribution. If an estimate (e.g. organisms in a suspension, radioactivity of a specimen) is repeated many times we can plot the values of the different estimates (*number* in this figure) against the number of times (*frequency*) that each particular value was recorded. The outcome is very often a bell-shaped curve as shown here.

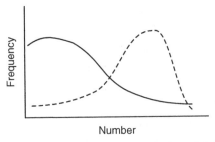

Fig. 6.2 Positively and negatively skewed distributions. With a positive skew, (solid curve) there is a tail towards the right (higher numbers), and with a negative skew (dashed curve) there is a tail to the left (lower numbers).

who is a poor starter: 0; 0; 1; 7; 9; 15; 30; 58; 95; 130; mean 34.50; median 12. (A bit like Keith Miller, though he was a *great* bowler too, *and* my favourite cricketer.)

That the **distribution is not normal** must be suspected whenever **the mean and the median are widely different**. Non-normal distributions are said to be skewed, positively or negatively (Fig. 6.2)

Normal distributions can have differing bell shapes, broad or narrow, but there is only one normal curve for a given mean (μ) and a given standard deviation (s). The most commonly used normal curve has $\mu = 0$ and $s = 1$. This is called the standard normal curve. The equation for this curve is:

$$y = \frac{1}{\sqrt{2\pi}}\, e^{-\frac{1}{2}x^2}$$

where x is the number of standard deviations and $e = 2.718\,28$ (see Chapter 5). Integration of this equation allows calculation of the area beneath the standard normal curve between any desired limits (Fig. 6.3 and Fig. 6.4).

6.5   Comparison of two means

This is one of the most important topics in statistics, and is much more complicated than a newcomer might expect. Two means *must* be different from each other (with at least 95% probability) when their 95% confidence limits do not overlap. For instance, two means of 15 ± 5 and 30 ± 6 are different with ≥95% probability; because there is only a 2.5% probability that the first mean is really greater than 20, and a similar probability that the second is less than 24.

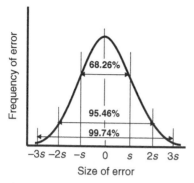

Fig. 6.3 Relation between standard deviation and areas under the normal standard curve. Only 4.54% of all errors will be more than 2 standard deviations (*s*) from the mean (0 in the figure), and only 0.26% will be more than 3 standard deviations away.

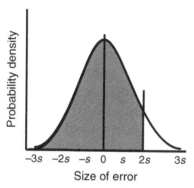

Fig. 6.4 Area under standard normal curve. Statistical tables give the area under the curve to the left (shaded) of a given positive standard deviation (*s*). This area is expressed in the table as a decimal fraction of 1 (which is the total area under the whole curve). When the size of error is + 1.96 × *s* then 95% of the area is to the left.

The rationale of tests to find whether two means are significantly different is first to suppose that the two means are not really different at all, and that many estimates of the two means would together generate a normal Gaussian curve centred on the assumed common mean. This is done by making a pooled estimate of the variance of this assumed common mean by pooling the actual variances of the two means that are being tested. Then the observed difference of the two means is expressed as a number of standard deviations from the supposed common mean. The probability that such a difference might have arisen by chance can finally be determined. All these manoeuvres are done for you by the equations which follow.

When the absolute value of the difference of two means, $|\bar{x}_1 - \bar{x}_2|$, and the standard deviations (s_1 and s_2) of each mean have been estimated from values of n_1 and n_2 that are **both** $\geq$**30** (though n_1 need not equal n_2) then the equation:

$$z = |\bar{x}_1 - \bar{x}_2|/\sqrt{[(s_1{}^2/n_1) + (s_2{}^2/n_2)]}$$

may be used, to decide whether the means are probably different, provided that s_1 and s_2 are close. The means are different, with $\geq$95% probability, if the absolute value of z is $\geq$2.24 (i.e. less than 2.5% probability that n_1 is not really greater than n_2 and less than 2.5% probability that n_1 is not really smaller than n_2).

When n_1 or n_2 is less than 30 then a different equation must be used:

$$t = |\bar{x}_1 - \bar{x}_2|/s_p\sqrt{[(1/n_1) + (1/n_2)]}$$

where s_p is the square root of this expression: $[(n_1 - 1)s_1{}^2 + (n_2 - 1)s_2{}^2]/(n_1 + n_2 - 2)$. Again, n_1 and n_2 need not be equal. The lowest value for t that is consistent with a 95% probability of the two means being different must be looked up in statistical tables in relation to the value of $n_1 + n_2 - 2$. For example, if n_1 and n_2 both were 3, then t would be 2.78 (for 95% probability), 3.50 (for 97.5% probability) or 4.60 (for 99% probability). If n_1 and n_2 both were 5, then t would be 2.31 (for 95% probability), 2.75 (for 97.5% probability) or 3.36 (for 99% probability). If the value of t (calculated from the equation) is $\geq$ the appropriate value of t found from the table, then the two means are probably different.

This **t-test** is one of the most commonly used statistical methods, and a basic tool in research. However, **a limitation of this test**, and of the large samples method, is that s_1 and s_2 must not be too different. The **F-test** can be used to decide whether s_1 and s_2 have values that are acceptable, but describing how to do the test goes beyond the scope of this chapter.

If you have only a small difference between means and only a small number of samples then it may not be possible to show that the difference is significant. However, if you had more samples (perhaps with smaller variances too) then the same small difference of means might prove to be significant.

Let's now show how these standard methods can be applied.

Suppose we grow a bacterial culture (in a minimal medium with no added amino acids) into late exponential phase, then make an extract of these organisms and assay it for the enzyme aspartokinase. We make four more extracts from four further cultures of the same organisms, grown and harvested in the same way, and we assay for aspartokinase in each extract.

The five results are (nmoles min^{-1} (mg protein)$^{-1}$): 29, 36, 31, 38, 30.

The mean is 32.8 nmoles min^{-1} (mg protein)$^{-1}$ and the median is 31 nmoles min^{-1} (mg protein)$^{-1}$.

These results ($n = 5$) are only a sample of very many similar estimates of aspartokinase that could be made, hence the sample ($n - 1$) standard deviation is 3.96 nmoles min^{-1} (mg protein)$^{-1}$. To find the 95% confidence limits of the mean we see from statistical tables that when $n - 1$ is 4 (as here) then the required value of t is 3.50, thus the confidence limits are $\pm$ 3.50 $\times$ 3.96 / $\sqrt{5}$ = $\pm$ 6.20.

Now we grow five cultures in minimal medium plus the amino acid L-methionine (100 mg L^{-1}), and make extracts and assay for aspartokinase as before.

The five results are (nmoles min^{-1} (mg protein)$^{-1}$): 27, 21, 23, 30, 28

The mean is 25.8 nmoles min^{-1} (mg protein)$^{-1}$ and the median is 27 nmoles min^{-1} (mg protein)$^{-1}$

The standard deviation ($n - 1$) is 3.70 nmoles min^{-1} (mg protein)$^{-1}$

The confidence limits are $\pm$ 3.50 $\times$ 3.70 / $\sqrt{5}$ = $\pm$ 5.79

In minimal medium the activity of aspartokinase was 32.8 $\pm$ 6.20 nmoles min^{-1} (mg protein)$^{-1}$

In medium plus methionine the activity was 25.8 $\pm$ 5.79 nmoles min^{-1} (mg protein)$^{-1}$.

Are these two means likely to be really different? The first mean (32.8) *could possibly* have a true value as low as 32.8 − 6.2 = 26.6, and the second mean *could* be as high as 25.8 + 5.79 = 31.59. We need, therefore, to test the probability that the mean in minimal medium is higher than the mean in medium plus methionine. This is a **one-tailed test**. Because n_1 and n_2 are each only 5 doing this leads to a slightly formidable calculation from the equation given above:

$$t = |\bar{x}_1 - \bar{x}_2| / [s_p \sqrt{(1/n_1 + 1/n_2)}]$$

where s_p is the square root of $[(n_1 - 1)s_1^2 + (n_2 - 1)s_2^2]/(n_1 + n_2 - 2)$

Work out s_p first: $=$ square root of
$$[(4 \times 3.96^2) + (4 \times 3.70^2)]/(5 + 5 - 2)$$
$$= \sqrt{[(62.79 + 54.79)/8]}$$
$$= \sqrt{14.70} = \mathbf{3.83}$$
Next evaluate $\sqrt{[(1/n_1) + (1/n_2)]} = \sqrt{(1/5 + 1/5)}$
$$= \sqrt{0.4} = 0.632$$
$$\text{Now}, t = (32.8 - 25.8)/(3.83 \times 0.632)$$
$$= 7.0/2.42 = \mathbf{2.89}$$

When $n_1 + n_2 - 2 = 8$ then the difference between means is real at 95% probability if $t \geq 2.306$, and at 99% probability if $t \geq 3.355$. Since $t = 2.89$ we can say that it is more than 95% (but less than 99%) probable that the mean from minimal medium is higher. Growth in the presence of methionine has very probably repressed aspartokinase in these bacteria.

Suppose that only three of the extracts grown without methionine, and three of the extracts grown with methionine had been prepared and assayed for aspartokinase, with these results:

No methionine: 29, 36, 31 nmoles min^{-1} (mg protein)$^{-1}$
Plus methionine: 27, 21, 23 nmoles min^{-1} (mg protein)$^{-1}$

Use the t-test to decide whether one of the means (32.00) is significantly higher than the other (23.67). When $n_1 + n_2 - 2 = 4$ the t value is 2.78 for 95% probability of difference. (There is no need to find the confidence limits.)

6.6   One-tailed and two-tailed tests of probability

A **one-tailed test** is used to establish whether **a larger mean 1 is significantly greater than a smaller mean 2**. The value of the error of the difference of the means will fall by chance within the small area to the right (that makes up only 5% of the total area under the bell-shaped curve) in only ≤5% of all trials (Fig. 6.5). Hence, when the error of the difference does lie in this small area we can say that there is a ≥95% probability that mean 1 > mean 2.

There is also a ≤5% probability that mean 1 could be significantly smaller than mean 2 because by chance the error of the difference of the means will

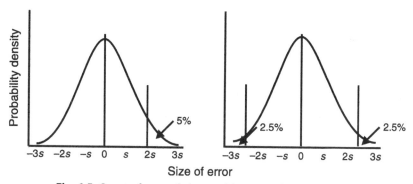

Fig. 6.5 One- and two-tailed tests of the error of the difference of two means.

fall within the area to the left that makes up 5% of the total area in ≤5% of all trials. The sum of these two probabilities could be >5%, in which case the probability of a true difference of the means would be <95%. A **two-tailed test** is therefore used to determine whether the **two means are significantly different** regardless of which appears the larger. When the error of the difference of the means falls within an area to the right that makes up ≤2.5% of the total area then the sum of the right and left probabilities must be ≤5%. In a two-tailed test, therefore, the t-test result is compared with the 2.5% value from statistical tables to try to establish that mean 1 ≠ mean 2 with 95% probability.

The values of s_1 and s_2 were very similar in the above worked-out example, so that the use of the t-test was justified. What if the s_1 and s_2 values had not been compatible? What if the distribution of results were not normal?

6.7   Non-parametric tests

These are statistical methods to compare means that do not require so many assumptions about the underlying (population) distributions. The distributions of the two populations do not need to be normal in non-parametric tests, but the distributions must be similar. The methods are often easier to explain and the calculations are frequently simpler than the standard tests that we have so far considered.

The **Wilcoxon (or Mann–Whitney) test** is a non-parametric alternative to the small-sample t-test for differences of means, and **can be used even when s_1 and s_2 values are not compatible,** although there is in fact no need to know the s_1 and s_2 values before doing the Wilcoxon test.

To illustrate the procedure we will use the five extracts (as before) from organisms grown without methionine, and the five extracts grown plus methionine, and assayed for aspartokinase. Results were (nmoles min^{-1} (mg protein)$^{-1}$):

Without methionine: 29, 36, 31, 38, 30, mean 32.80
Plus methionine: 27, 21, 23, 30, 28, mean 25.80

The first step is to pool all these ten results, as if they all came from one set of samples, and put them in ascending order of magnitude.

Results	21	23	27	28	29	30	30	31	36	38
Rank	1	2	3	4	5	6.5	6.5	8	9	10
m/-	m	m	m	m	–	m	–	–	–	–

We assign the data in this order the ranks 1, 2, 3, ... , 10 and indicate, for each rank number, whether it came from the no methionine (–) or the plus methionine (m) series. When two (or more) results are the same (as are 30 and 30 here) they are all given the same averaged rank number (here the average of 6 and 7).

Next, add the rank numbers for the no methionine results:

$$5 + 6.5 + 8 + 9 + 10 = 38.5 = W_1$$

Then add the rank numbers for the plus methionine results:

$$1 + 2 + 3 + 4 + 6.5 = 16.5 = W_2$$

($W_1 + W_2$ *must* always equal the sum of all the integers from 1 to $(n_1 + n_2)$ inclusive; otherwise you have made a mistake.)

Now work out $U_1 = W_1 - [n_1(n_1 + 1)] / 2 = 38.5 - (5 \times 6) / 2$
$$= 38.5 - 15 = 23.5$$

(n_1 is the number of values (5) in the no methionine set)
and similarly evaluate $U_2 = W_2 - [n_2(n_2 + 1)] / 2 = 16.5 - 15 = 1.5$
(n_2 is the number of values (5) in the plus methionine set)
($U_1 + U_2$ *must* always equal $n_1 \times n_2$; otherwise there is an error.)
Finally we take whichever is the smaller, U_1 or U_2, and call this U. Thus, in our case $U = 1.5$.

For the no methionine mean to be higher with 95% probability our U value must be smaller than, or equal to, $U'_{0.05}$, which is a number, dependent on the values of n_1 and n_2, found from statistical tables. When n_1 and n_2

both are 5 then $U'_{0.05}$ is 2, while $U'_{0.01}$ (99% probability) is 0. Hence the no methionine mean is higher with more than 95% probability, but with less than 99% probability, the same result as that given by the t-test.

This U-test is not applicable if either n_1 or n_2 is less than 4, and when n_1 and n_2 are both greater than 10 a different ranking test is better to use. A small problem is that when values of n_1 and n_2 each get up towards 20 the ranking procedure becomes decidedly awkward to get right. In other respects the non-parametric test is easier. However, before doing the calculation you do need to know all the values that contributed to each mean, whereas in the t-test the two means and their standard deviations are enough.

6.8 Analysis of variance (ANOVA)

Up to now we have considered tests to compare two means. What if you want to test three or more means to find whether they are equal or not? You could test the means in pairs: does mean 1 = mean 2?; does mean 2 = mean 3?; does mean 1 = mean 3? What if you found that mean 1 = mean 2 and mean 2 = mean 3 but mean 1 ≠ mean 3? No useful conclusion could be reached. Testing such binary comparisons soon becomes excessively laborious as the number of means rises.

What you need is a single procedure to decide whether or not all of several means are equal. This procedure does exist, and is called **ANOVA** (**AN**alysis **O**f **VA**riance). Describing how to do an ANOVA test and the prerequisite conditions would make this chapter very long. Good explanations are easy to find (see Further reading).

Statistics is a degree-level subject, and you must realise that this chapter is no more than a very simple guide. More reading is essential if you want a fuller understanding. Many elementary texts are available. Use whatever you can get from your library (or the Internet).

Be very careful when you use a statistical method to test results of your own. Using the right method (for your data) in the right way is not always a simple matter. It is all too easy to get an answer that looks good but which does not stand up to critical inspection.

7 | Preparing solutions and media

Round about the cauldron go;
In the poison'd entrails throw.

William Shakespeare

7.1   Amounts and concentrations

Many students have great difficulty in distinguishing clearly between amounts of a substance, and the concentration of a substance in a solution, or the proportion of a substance in a mixture.

An amount is a definite quantity of a solid, liquid or gas. Examples are: 10 mg anhydrous potassium nitrate; 1 g of 10M sulphuric acid; 25 ml of 2M sulphuric acid; 10 L hydrogen (at STP); 3 ml of 0.9% (w/v) sodium chloride in water; 2×10^7 staphylococci; 5 mmoles ammonium sulphate.

The relative molecular mass (abbreviated RMM or mol. wt) of a substance, expressed in grams, is one mole of that substance. A mole (abbreviated mol) is a definite quantity of a substance, even if the RMM of that substance is not known, because **the number of molecules in one mole** of any pure substance is always **6.02×10^{23} (Avogadro's number)**. This is not the same as the number of atoms in a molecule; for example oxygen or hydrogen gases each have 2 atoms per molecule, methane has 5 atoms per molecule while glucose has 24 atoms in its molecule.

A concentration is an expression of proportion, whether this is for example the strength of a solution or the percentage of an organism that is nitrogen. Knowing a concentration is very important, but the concentration, of itself, does not tell the amount of material. Concentrations of solutions are quoted in a variety of ways. It is important to know what all of these mean, and how to make up solutions so described. Unless it is stated otherwise, the solvent will always be water.

Examples are: 5 g ammonium chloride per litre; 2 millimolar ATP; 1 µmol NADPH per 5 ml of stock solution; 3% (v / v) acetonitrile in water.

7.2 Expressions of concentration

Weight per unit volume of solution

Examples of weight per unit volume of solution are: 12 g glucose L^{-1}; 50 µg penicillin ml^{-1}.

The meaning is clear, but it is necessary to think about the total volume of solution that is actually required. Perhaps only 50 ml is needed of a solution described as 12 g glucose L^{-1} so that only 0.6 g glucose ($12 \times 50 / 1000$) is actually weighed out. It is not always simple to decide how to prepare a small volume (say 10 ml) of a dilute solution, such as 5 µg of biotin ml^{-1}. In such a case, it is not feasible to weigh 50 µg of biotin. What has to be done is to take a quantity of biotin that can be accurately weighed, perhaps 10 mg, and dissolve this in (say) 10 ml of water to give a solution that contains 1 mg (1000 µg) biotin ml^{-1}; then make up 50 µl of this ($\equiv$ 50 µg biotin) to 10 ml with water (50 µg in 10 ml $\equiv$ 5 µg ml^{-1}) to produce the required volume of solution at the required concentration. The stronger solution may be kept as a stock for future use, provided that it is stable under the conditions of storage.

Remember that always **weight per unit volume of solution, not of solvent** is to be understood. When making strong solutions (e.g. 200 g sucrose L^{-1}) it is essential not to use too much solvent initially. If 1 litre of water had been used to dissolve 200 g sucrose the final volume of solution would be considerably more than 1000 ml. The correct procedure in this example is to take about 700 ml of water, dissolve the sucrose and then make up the volume to 1 litre with water.

Even when a concentration is given in weight per volume terms it might be necessary to make a calculation from the RMM of the solute. For instance, account may have to be taken of water of crystallisation. Thus, a solution may be described as containing 12 g anhydrous sodium acetate L^{-1}. If only sodium acetate trihydrate were available in your laboratory then the greater amount to be weighed to make a litre of solution containing the correct amount of sodium acetate would have to be worked out:

CH_3COONa, mol. wt $= 82$; $CH_3COONa.3H_2O$, mol. wt $= 136$

Hence, instead of 12 g of anhydrous compound, the amount of the trihydrate must be:

$$12 \times 136/82 \ = \ 19.9 \text{ g sodium acetate trihydrate}$$

If instead you had to make a solution described as containing 12 g sodium acetate trihydrate L^{-1} and only the anhydrous material were available, then the amount to be weighed would be:

$$12 \times 82/136 = 7.2 \text{ g anhydrous sodium acetate}$$

Always remember that in going from a less hydrated to a more hydrated form, the amount needed will increase, while in going from a more hydrated to a less hydrated form the amount needed will decrease.

Molarity

Examples of molarity are: 0.1 M KCl; 1 μmole NAD^+ ml^{-1}.

A molar solution (1 M) contains 1 mole of solute per litre of solution. The only accepted abbreviation of mole is mol, and the abbreviation for molar is M. Much confusion arises from a failure to recognise that **expressions such as 0.2 M, 3 mM or 5 μM are all concentrations** $\equiv$ 0.2 mol L^{-1}; 3 mmol L^{-1}; 5 μmol L^{-1}) **and not amounts.** To write for instance '4 mM per 100 ml' is a very common kind of error, and it comes from the mistaken idea that M can be an abbreviation for mole. What should be written is 4 mmol per 100 ml (or 40 mM, which is the same concentration). A 1 M solution is 1 M whether you have 10 ml, 500 ml or 10 L of it. However, in 10 ml (of 1 M solution) only 0.01 mole of solute is present, whereas in 10 L there are 10 moles.

The same concentration can be expressed in several ways: 0.1 M NaCl is the same as 100 mM NaCl and this is the same as 0.1 mmol ml^{-1} or 100 μmol ml^{-1}. Which of these is to be preferred will usually depend on the context, though the most compact form (0.1 M NaCl in these examples) should most often be the choice.

To prepare a given volume of a solution of known molarity requires a simple calculation to determine how much of the substance must be weighed. For example, to make 500 ml of 0.2 M $Na_2HPO_4.12H_2O$ (1 mole = 358 g) one would need 358 / 2 (500 ml needed rather than 1 L) ÷ 5 (0.2 M rather than 1 M) = 35.8 g. This should be dissolved in about 400 ml of water and then made up to 500 ml with more water.

Difficulties in making small volumes of weak solutions are handled as described above – make a stronger solution and dilute it appropriately.

Normality

Examples of normality are: 2 N NaOH; 6 N HCl.

This term is obsolete. It comes from a time when titrations of acids and bases were much more widely used than they are now. A normal solution of an acid (1 N) contains 1 g of **ionisable** hydrogen per litre. Hence 1 N H_2SO_4 is 0.5 M; 1 N H_3PO_4 is 0.33 M; 1 N HCl is 1 M. A 1 N solution of a base will neutralise an equal volume of 1 N acid. Hence 1 N NaOH is 1 M. The amount of a substance needed to make 1 L of a 1 N solution is called a gram equivalent of that substance. The gram equivalent can in many cases be the same as the mole (e.g. HNO_3; NH_4OH). Solutions of oxidising and reducing agents can also be described in terms of normality. A gram equivalent is the weight that accepts one mole of electrons (oxidising agent) or that donates one mole of electrons (reducing agent).

Percentage

Examples of concentrations given as percentages are: 0.9% (w / v) NaCl; 10% (v / v) ethanol.

Percentage weight / volume (w / v) means the weight (in grams) of a solute present in 100 ml of a solution. Hence 0.1% (w / v) sodium azide means 0.1 g (100 mg) of sodium azide dissolved in water and made up to 100 ml. **The molecular weight of the solute does not come into consideration while making a solution that is defined by percentage.**

Percentage volume / volume (v / v) means the volume (in ml) of a liquid solute present in 100 ml of solution. Hence 5% (v / v) methanol means 5 ml of methanol made up to 100 ml with water.

Percentage weight / weight (w / w) means the weight (in grams) of a substance present in 100 g of a material or solution. Using % (w / w) to describe a solution is extremely uncommon. Generally, % (w / w) is used to express the proportion that a compound or element represents of the total weight of a material. We can say for instance that carbon makes up 40%

(w / w) of glucose, or that DNA is about 1% (w / w) of a bacterium. Be very careful not to get confused here. You might for instance be told that nitrogen makes up 15% (w / w) of the dry weight of an organism, and that protein makes up 35% (w / w). These numbers, though both may be correct, do not mean that 50% of the weight is accounted for! Remember that the protein will itself contain much of the nitrogen (and will also contain carbon, hydrogen, oxygen and sulphur).

7.3   Solutions containing a mixture of solutes

More often than not, the solutions used in biochemistry contain several different substances dissolved together in water. Quite complicated solutions are needed for enzymic and spectrophotometric assays, for electrophoresis, or for genetic manipulations. Some growth media are particularly elaborate solutions.

The majority of these solutions are buffers, in which various reagents and/or nutrients are dissolved. A buffer is a solution to which the addition of a limited quantity of acid or alkali has a very small effect on the pH value of the solution. Most often, a buffer is a solution containing a weak acid and its anion (e.g. $H_2PO_4^-$ and HPO_4^{2-}; CH_3COOH and CH_3COO^-) at similar (but not necessarily equal) concentrations, or else a weak base and its cation, e.g. Tris, which is tris hydroxymethyl aminomethane [$(CH_2OH)_3CNH_2$] and [$(CH_2OH)_3CNH_3^+$] again at similar concentrations. The higher the molarity of the components of a buffer in a solution, the more acid or alkali can be added before there is an appreciable change in the pH value of the solution. Any given buffer combination is effective over only a limited range of pH values, about 1 unit each side of the value for optimal buffering. This optimal value is determined by the pK_a value of a weak acid or the pK_b value of a weak base. Hence, acetate buffer is useful in the range pH 3 to 5, phosphate at pH 6 to 8 and Tris at pH 7 to 9. Unfortunately, many of the commonly used (and cheap) buffers are metabolites themselves (acetate, phosphate, bicarbonate) or may be inhibitors of some enzymes (Tris) or may change their pH value with change of temperature (Tris). A buffer should ideally be freely soluble in water, non-metabolisable and non-inhibitory. Such buffers are readily available (e.g. HEPES (4-(2-hydroxymethyl)-1-piperazineethane sulphonic acid), pK_a 7.55; MOPS (4-morpholinepropane sulphonic acid), pK_a 7.2; TES (N-(tris (hydroxymethyl)methyl)-2-aminoethane sulphonic acid), pK_a 7.4), but they are not cheap, so that less satisfactory buffers, especially phosphate and Tris, are still very widely used.

It is perhaps unfortunate that in the microbiological literature the conventional way of describing a growth medium is terse, being no more than a list of the quantity of each component that is present in 1 L of medium (when this medium is at the concentration that is actually used, rather than at the higher concentration that might be stored). These descriptions are by no means easy for a novice to convert into a working set of instructions, or recipe, for producing the medium. In particular, the descriptions may not always reveal that many of the components are added as concentrated solutions, rather than as solids, or that the medium is normally made up at double its final working strength (i.e. listed components for 1 L dissolved to make only 500 ml) so that optional extra solutions can be added without making the medium too dilute, and to save space during storage.

7.4   Solutions containing a mixture of solvents

Mixtures of two or more different liquids are often used, especially as solvents for liquid chromatography. The best way of describing these is to state the proportions in which the constituents were mixed, as for instance: methanol / pyridine / water (40 + 10 + 10 by volume) because the total volume of the mixture (and hence the final % (v / v) of each component) does not necessarily equal exactly the sum of the separate volumes of the three liquids.

When solutes are described as being in (for example) 10% (v / v) aqueous ethanol, the usual way of preparing the solution is first to make up a suitable quantity of the 10% (v / v) ethanol (say 10 ml ethanol + 90 ml water) and then to dissolve the solutes in this liquid.

7.5   Dilutions

The same degree of dilution of a solution (e.g. 10-fold) can be described as 1 / 10; 1 : 10 or 1 ml + 9 ml. Only the last of these is unambiguous. The meaning of **1 / 10 and of 1 : 10 ought to be 1 vol. made up to 10 vol.** (which is very nearly the same as 1 + 9) but unless you are sure about the author's usage there may be uncertainty whether or not 1 vol. plus 10 vol. is the intended meaning. Both 1 / 10 and 1 : 10 indicate proportions, and, when used correctly could equally well mean that 10 ml was made up to 100 ml, or that 1 ml was made up to 10 ml, or that 5 ml was made up to 50 ml. However, the 1 + 9 format does generally have units of volume attached, as for example 100 μl + 9.9 ml. In a case such as this, the degree of dilution (here 100-fold) may not be immediately obvious; it is necessary to find the ratio of the total volume (10 ml) to the volume of the solution that is being diluted (0.1 ml).

Sometimes one has to determine what dilutions are needed to produce a solution of a desired concentration from another stronger solution. Some examples follow to show how this is done.

(1) Given a solution containing 7 mg X ml^{-1}, make a solution (10 ml is sufficient) containing 2 mg X ml^{-1}. The dilution required is 2 / 7 (because 2 ml of the stronger solution will contain 14 mg X and so, when made to 7 ml, the concentration will become 2 mg X ml^{-1}). Since at least 10 ml of the diluted solution is wanted, an appropriate procedure is a 4 / 14 dilution, 4 ml of the stronger solution + 10 ml water.

(2) Given a solution containing 10 mg NaCl ml^{-1}, make a 0.1 M solution (50 ml is sufficient). In this case it is first necessary to express the concentrations of both solutions in the same units, either as mg ml^{-1} or as molarities. Here we shall do both, to show that each method gives the same answer.

The mol. wt of NaCl is 58.5, and therefore 10 mg ml^{-1} (≡ 10 g L^{-1}) is (10 / 58.5) M = 0.171 M. The required dilution is thus 0.1 / 0.171 or 1 / 1.71 or 1 + 0.71.

A 0.1 M solution of NaCl contains 5.85 g L^{-1}, or 5.85 mg ml^{-1}. The required dilution is thus 5.85 / 10 or 1 / 1.71, as found before. This is the same degree of dilution as 10 / 17.1, but 17.1 ml of 0.1 M NaCl is not enough, we need 50 ml. Consequently, 30 / 51.3 is appropriate: take 30 ml of the stronger solution and add 21.3 ml water.

(3) Given a 10% (w / v) solution of glucose, make at least 20 ml of a solution containing 1 μmole ml^{-1}. Again the concentrations of both solutions must first be expressed in the same units. The stronger solution contains 10 g glucose in 100 ml solution ≡ 100 g L^{-1}. The mol. wt of glucose is 180 and so the stronger solution is (100 / 180) M = 0.556 M. This means 0.556 moles L^{-1}, which is the same as 0.556 mmoles ml^{-1} or 556 μmoles ml^{-1}. The required dilution is therefore 1 / 556. As 556 ml is

much more than is needed, an appropriate procedure would be to take 0.1 ml of the stronger solution and add 55.5 ml water.

In calculating dilutions bear in mind the realities of working in a laboratory. It is not possible to dispense volumes that include fractions of a microlitre, nor is it easy to find a container for >10 litres of a liquid, nor to lift and mix such big volumes.

Serial dilution

Serial dilution is the preferable alternative to making single-step dilutions of very small volumes of a strong solution or to taking very large volumes of the diluent. The principle is simple; if a solution is diluted sequentially (e.g. dilute solution A 1 / 100 to give solution B, then dilute solution B 1 / 100 to give solution C, then dilute solution C 1 / 100 to give solution D) the final degree of dilution (of solution D in the example) is the product of the successive dilutions (1 / 1 000 000 in the example). A dilution of 1 / 556 (example 3 above) could be made: 1 ml 10% (w / v) glucose + 4.56 ml water (1 / 5.56) followed by 0.3 ml (of the first dilution) + 29.7 ml water (1 / 100 dilution) to give 30 ml of diluted solution of the required strength (1 μmole ml^{-1}).

The precision of dilutions

When two chemically different but miscible liquids are put together, the volume of the mixture does not necessarily equal precisely the sum of the volumes of the two separate liquids. Combining exactly 50 ml of ethanol and exactly 50 ml of water does not give exactly 100 ml of mixture, though the discrepancy is small. In the same way, the result of mixing together a strong aqueous solution and pure water is not strictly additive. The dilutions made in routine biochemical work are not therefore always scrupulously accurate, even if the correct volumes of liquids being mixed are dispensed accurately. In practice these volumes are not usually measured with great care, and the degree of dilution only approximates to what was intended. The errors in dilutions are, however, acceptably small (provided that no mistakes were made in the calculation or the dispensing) for all but the most fastidious experiments.

In working out results of experiments, **errors in calculation are much, much more common than errors in dilution**, though students are far readier to attribute mistakes to 'pipetting errors' than to an inability to do the sums properly.

7.6   Making solutions from impure solutes

Sometimes a solution has to be made from a material which is known to be impure. In general, the procedure is to make a relatively strong solution of the impure material, determine the concentration of the desired solute in this solution, and then dilute it appropriately. Finding a suitable method to determine the concentration is often the hard part!

For example, ATP of 100% purity cannot be bought. Nevertheless, one may need to prepare a solution of ATP at a definite concentration, such as 1 mM. To do this, take a weighed amount of the ATP and make a solution (say 50 ml) that is very likely (from an estimate of the probable purity of the ATP) to be stronger than what is finally wanted. Then dilute a sample of this solution (by a known factor) into buffer of a known pH value, and measure its absorption spectrum in the region 220 to 290 nm. Provided that the spectrum matches that of ATP at the particular pH value chosen, then no UV absorbing impurities are likely to be present, and so the concentration of ATP in the diluted solution can be calculated from its maximum absorbance and the extinction coefficient of ATP at that same wavelength and pH value. Now the concentration of the undiluted solution becomes known (suppose it happens to be 1.23 mM) and this solution can now be diluted to make it accurately 1 mM (10 ml + 2.3 ml water). To establish the % (w / w) purity of the ATP, the total amount of ATP found in the solution before dilution (0.0615 mmoles in 50 ml), expressed as a weight of ATP, can be compared to the amount of impure ATP that was actually weighed.

As a check on the reliability of the solution of 1 mM ATP one could measure the concentration of total phosphorus in the solution, which should be 3 mM. An appreciably lower concentration of phosphorus would be very worrying, but a slightly higher value would not, because some of the impurities in the ATP might also contain phosphorus and contribute to the total. Although the final solution may be accurately 1 mM ATP, remember that other unknown materials are also present, and these may, or may not (one hopes), interfere with the intended function of the ATP.

8 | Enzymes

Was this the face that launch'd a thousand ships,
And burnt the topless towers of Ilium?

Christopher Marlowe

An enzyme is a catalyst produced by a biological process. Very nearly all known enzymes are proteins, but only rarely can one detect an enzyme by identifying the enzymic protein in a mixture. Usually the presence of an enzyme is revealed by showing its catalytic effect on a reaction.

This is done by comparing the rate of the reaction with and without the material that might contain the enzyme. An assay system has to be made which, when complete, will contain: a buffer to give a pH value that is optimal for the enzyme; the substrate of the reaction; any necessary cofactors and activators; and the test material. The mixture without the substrate, or without the test material, is brought to the desired temperature, and then the reaction is started by adding the substrate or test material. There must also be a method to assess the rate of the expected reaction. Such a method could be, for instance, uptake of oxygen; reduction of a cofactor; appearance of the product of the catalysed reaction; release of inorganic phosphate. An observed catalytic effect can be shown to be enzymic by demonstrating that the effect is lost when the presumed enzyme is denatured, as by heat.

The rate of reaction in the complete system must always be compared to the rate in the assay system lacking the test material (**no enzyme control**), and to the rate in the system lacking the substrate (**no substrate control**). The purpose of the first of these controls is obvious – if the enzyme is present in the test material then the reaction must be faster in the complete system than in this control. The no substrate control is equally important, but rather more difficult to understand. The test material might contain other enzymes (and substrates if the material is impure) than the one which is intended to be measured. Thus, for example, a detected uptake of oxygen

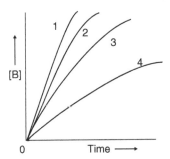

Fig. 8.1 The initial rate of the enzyme-catalysed reaction A → B + C is determined by measuring the concentration of B at various times after adding the enzyme. Curve 1 shows the rate with a high concentration of substrate. Halving the concentration of substrate (though using the same amount of enzyme) leads to a lower initial rate of reaction (curve 2). The rate becomes still lower when the concentration of A is decreased further (curves 3 and 4), while the amount of enzyme used is unaltered.

could be attributed to the expected enzyme only if the rate of uptake in the no substrate control were slower than in the complete system.

8.1   Kinetics of enzyme-catalysed reactions

The kinetics of enzyme-catalysed reactions is a large and complex subject. Here a very simplified discussion is all that will be attempted.

The rate of an enzymic reaction is dependent on the concentration of the substrate when other conditions (optimal pH value and temperature, amount of enzyme) are constant. If the molecules of substrate are in great excess over the molecules of enzyme then the reaction proceeds at the maximum rate (V_{max}) that is possible (with the amount of enzyme that is present).

When the substrate is in excess one can imagine a queue of substrate molecules waiting for an unoccupied molecule of the enzyme to become available. As the reaction continues the substrate becomes depleted, so that a state is reached where the enzyme must wait for a molecule of substrate to approach. Hence, a typical enzymic reaction has a constant rate for some period after its start, but gradually the rate decreases, and will be zero when all the substrate has been used or equilibrium is reached. The rate may decrease for another reason too; that is, the enzyme may not be stable under the conditions of assay.

Having done an experiment like that shown in Fig. 8.1, we can then plot the initial rate of the enzymic reaction against the concentration of substrate (Fig. 8.2).

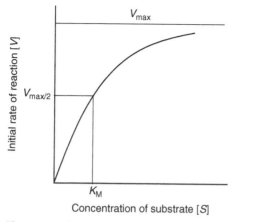

Fig. 8.2 As the concentration of substrate increases the initial rate of an enzymic reaction approaches a maximum (for the amount of enzyme used) that is called V_{max} and the concentration of substrate that gives half this maximum rate is called the Michaelis constant (K_M) of the enzyme.

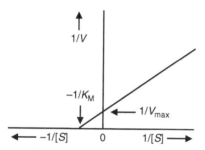

Fig. 8.3 The Lineweaver–Burk plot.

The Michaelis–Menten equation:

$$V = V_{max} \times \frac{[S]}{[S] + K_M}$$

shows the relation between V, $[S]$, V_{max} and K_M.

It is not always easy, from the kind of plot shown in Fig. 8.2, to establish accurately what is the value of V_{max}, and hence of $V_{max}/2$ and so K_M. Plotting the reciprocals of the initial velocities against the reciprocals of the corresponding concentrations of substrate (the Lineweaver–Burk plot) very often will give a straight line from which K_M can readily be found (Fig. 8.3). Rearranging the above equation shows why this works:

$$1/V = ([S] + K_M)/V_{max}[S]$$
$$= [S]/V_{max}[S] + K_M/V_{max}[S]$$
$$= 1/V_{max} + (K_M/V_{max}) \times 1/[S]$$

Thus, plotting 1 / V against 1 / [S] gives a straight line of intercept 1 / V_{max} (on the 1 / V axis).

Furthermore, when $1/V = 0$, then

$$(K_M/V_{max}) \times 1/[S] = -1/V_{max} \text{ and so}$$
$$K_M \times 1/[S] = -1$$
$$1/[S] = -1/K_M \text{ when } 1/V = 0$$

i.e. the intercept of the line on the 1 / [S] axis = -1 / K_M.

The initial rate of an enzymic reaction is also proportional to the amount of enzyme that is present. Doubling the concentration of the enzyme should double the initial rate. This simple relation will no longer be true when so much enzyme is present that the substrate is not in excess, even at the onset. Neither will it apply if the initial rate becomes so fast that it cannot be measured accurately.

In summary, a satisfactory assay of an enzyme will show a steady initial rate, and the rates will be very much lower (or still better, zero) in the no enzyme and no substrate controls. The initial rate will be proportional (within limits) to the amount of enzyme used and to the concentration of the substrate.

By whatever means it has been detected, the initial rate should finally be expressed as µmoles of substrate used per minute (µmol min^{-1}). This number next can be related to the volume of enzyme solution that was used in the assay so as to calculate µmol min^{-1} (ml of enzyme solution)$^{-1}$. Measuring total protein in the enzyme solution (x mg ml^{-1}) then leads to the specific activity of the enzyme: µmol min^{-1} (mg protein)$^{-1}$. This is numerically the same as 'units of enzyme'.(mg protein)$^{-1}$ because one unit of an enzyme is defined as that quantity of enzyme which gives a rate of reaction of one µmol min^{-1}.

8.2   Turnover number

The **turnover number** (or catalytic constant) of an enzyme is the number of substrate molecules transformed per second by one molecule of the enzyme, at the optimal pH value and with the substrate in excess, so that V_{max} is achieved. The turnover number is given by the relation: units of enzyme × mol. wt of the enzyme × 10^{-3} / (60 × n), where n is the number of catalytically active sites per molecule of enzyme. To determine the turnover number the enzyme must be pure, and its mol. wt must also be known.

8.3 Extracting enzymes from microorganisms

Microorganisms release some enzymes (exoenzymes) into the surrounding medium. Other enzymes (endoenzymes) only occur inside the organisms that have made them. These latter enzymes are much the larger group, and a very common difficulty in assaying them is that a substrate may not penetrate the wall and membrane(s) of an intact organism, so that no enzymic activity is seen although the enzyme may really be present internally. Many methods have been developed to overcome this problem.

Boiling in 2 M NaOH is an effective way of dissolving the wall and membranes, and bringing all the cytoplasmic protein into solution, so that the total amount can be measured. However, all enzymic activity will be lost because the proteins are denatured. It is essential that an enzyme is extracted (or else made accessible to its substrate) by a method that preserves its catalytic action. Examples of such methods are:

(1) Grinding a thick suspension of the organisms with a finely powdered abrasive material, such as alumina (aluminium oxide). This destroys the structure of the organisms, and protein escapes into solution, but it is very difficult to achieve a breakage of more than about 20% of the organisms in the suspension.

(2) Adding a suspension of organisms to an excess of cold acetone (−18 °C) ruptures the membranes, and this enables many substrates to enter, though the internal proteins may be partly denatured by the solvent.

(3) Washing the organisms with a detergent (e.g. 0.3% (w / v) cetyl trimethylammonium bromide in water) also may damage membranes, and is less likely to denature proteins.

(4) Ultrasonic vibration will break organisms and release proteins into solution effectively. However, the released enzymes are not always active, perhaps because of the generation of free radicals that occurs during sonication.

(5) Liquid shear forces, achieved by the sudden decrease of a very high pressure, inside a robust steel apparatus (the French pressure cell or the Hughes' press) are very effective in breaking rod-shaped organisms without causing denaturation of enzymes. These devices are less successful with spherical organisms because these have much greater mechanical strength.

(6) Digestion of the wall with a lytic enzyme (e.g. lysozyme) followed by osmotic bursting of the protoplasts is a very gentle method. The difficulties lie in finding an effective lytic enzyme for a particular organism, and establishing the conditions in which lysis is most efficient.

The method to use is the one that works for its intended purpose! If an enzyme is to be purified, then it must be obtained in a soluble extract. The method of choice in such a case might therefore be the French pressure cell or lysis. If detection of an enzyme is the aim, then treatment with detergent may be adequate.

8.4   Coupled assays

To measure the rate of an enzymic reaction one must have a way of measuring either the disappearance of a substrate or the appearance of a product. Suppose we have a reaction:

$$A + H_2O \rightarrow B + C$$

and no simple method can measure A, B or C. This problem may be overcome if we have available a purified enzyme that has no action on A but produces a product from B or C that can be measured, thus for instance:

$$A + H_2O \rightarrow B + C \rightarrow B + X + Y$$

The added (coupling) enzyme has converted C to X and Y, where X or Y is readily detectable. The appearance of X will be a true measure of the rate of hydrolysis of A only if C is converted to X by the coupling enzyme very rapidly and quantitatively. For this to happen, the coupling enzyme must be added in large excess (enough to give at least a 100-fold greater rate of reaction with C than the rate of breakdown of A by the first enzyme) and must have a small K_M for C. Of course, the coupling enzyme must also itself have no reaction with A. An example of an assay with a coupling enzyme is that for diaminopimelate epimerase, which catalyses:

LL-diaminopimelate $\rightarrow$ *meso*-diaminopimelate

This reaction could be followed by the decrease in optical rotation while an optically active isomer (LL-) is converted into an inactive form (*meso-*), but rather a high concentration of LL-diaminopimelate is needed to give a measurable rotation, and a crude enzyme may contain other optically active compounds. Much better is to add in excess purified *meso*-diaminopimelate dehydrogenase, which catalyses

$$meso\text{-diaminopimelate} + NADP^+ + H_2O \rightarrow$$
$$\text{tetrahydrodipicolinate} + NH_3 + NADPH + H^+$$

and has no action on LL-diaminopimelate. The rate of increase in absorbance (caused by the formation of NADPH) is a measure of the rate of formation of *meso*-diaminopimelate from the LL-isomer.

8.5   Purifying enzymes

In order to study the properties and structure of an enzyme it must first be purified. To start, one must find a source of the enzyme from which the enzyme can be obtained as a soluble extract in good yield. Insoluble material is removed from the extract by centrifuging, and low molecular weight materials eliminated by dialysis or gel-filtration. The chief problem is to separate the desired enzyme from all other high molecular weight substances (principally proteins) that remain in the extract, without inactivating the required enzymic activity by inappropriate conditions. Over a period of almost 100 years many methods have been devised to try to achieve this separation. Very rarely can complete purification be reached by a single step. Choosing which methods to use, and in what sequence, is almost an art, and usually success is gained only after much trial and error. Published procedures seldom reveal all the steps that were tried and were unprofitable, and they may give a misleading impression that purification was easier than really it was.

Methods of purification

Two of the most important are **fractional precipitation** and **column chromatography**. Different proteins have different solubilities in aqueous

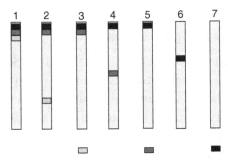

Fig. 8.4 Elution of protein from DEAE-cellulose with NaCl. (1) A mixture of three proteins is loaded onto a column. (2) The column is eluted with 0.1 M NaCl. (3) One of the proteins is eluted by this concentration of salt. (4) The column is eluted with 0.2 M NaCl. (5) A second protein is eluted. (6) The column is eluted with 0.3 M NaCl. (7) The third protein is now eluted.

solutions to which ammonium sulphate or an organic solvent such as acetone are added. If acetone is added to an aqueous extract to make 10% (v/v) acetone then some protein may precipitate. This is removed, and more acetone is added to give 20% (v/v). Now there may be a further precipitate, which is again removed. More additions of acetone should give other precipitates of protein. The separate precipitates are redissolved in aqueous buffer and each solution is assayed for the desired enzyme. The aim is to discard the precipitates that do not show enzymic activity. Fractionation with ammonium sulphate is done in the same way, by making successive additions of this salt and collecting the separate precipitates.

There are two kinds of column chromatography – ion-exchange and molecular exclusion. Proteins are charged molecules, and at an appropriate pH value their negative charges will cause them to bind to an anion-exchanger, such as DEAE-cellulose. Other cations (Na^+ or K^+) can displace the proteins: low concentrations of NaCl will dislodge weakly bound proteins, and different proteins will be removed as successively higher concentrations of NaCl are used (Fig. 8.4).

In molecular exclusion chromatography the sizes and shapes of different protein molecules are exploited to effect their separation. The column is packed with small beads of Sephadex (a modified carbohydrate) or polyacrylamide. The beads have minute pores of such a size that large proteins cannot enter, but small ones can, and these can also escape through the pores later. Consequently, the largest proteins travel most quickly down the column since they are not retarded by the beads (Fig. 8.5).

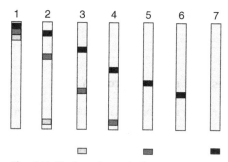

Fig. 8.5 Elution of protein from a Sephadex column with buffer. (1) A mixture of three proteins is loaded onto a column. (2) Elution with buffer causes the largest protein to move most rapidly down the column, and the smallest to move most slowly. (3) The largest protein emerges from the column. (4) to (7) Further elution leads first to the emergence of the protein of intermediate size (5) and then to the smallest protein (7).

Notice carefully that even when a protein begins to migrate as soon as elution starts this protein will not emerge until the buffer (in which the solid material of the column was at first suspended) has been displaced. This quantity of buffer is called the void – or interstitial – volume of the column.

At every step in purification one should measure the specific enzymic activity and the total volume of liquid in which the enzyme is present. From the specific activities at each step the degree of purification can be found (e.g. if the specific activity goes up fivefold as one step follows another, then the purification is also fivefold). By comparison of the total amount of enzyme present at each step the recovery of active enzyme is determined, e.g. if the total amount of enzyme at step 1 is 400 units and at step 2 is 350 units then the percentage recovery at step 2 is (350 / 400) ×100, which is approximately 88%. If after the next step 175 units remained, then this individual step would have given a recovery of 50%, and an overall recovery of 44%.

Ideally, purification should be continued until only a single protein (as assessed by every available method) with enzymic activity is left. This is not always possible owing to poor recoveries at some steps. Often less than 10% of the initial amount of enzyme remains by the last step in a purification. The final purification factor is not itself an indicator of the purity of the enzyme: if an enzyme makes up only 0.1% of the total protein in an extract, then a 100-fold purification will not be complete, whereas if an enzyme makes up 10% of the total cellular protein then a 10-fold purification should be complete.

8.6   Enzymic activities and rates of growth

An enzyme that plays a vital role in metabolism is sometimes found with an activity in an extract that seems surprisingly low. This may well be a consequence of poor extraction or unsatisfactory assay, but before deciding that there is such a problem one should calculate what is the lowest activity of the enzyme that is consistent with the observed rate of growth. In these calculations some estimates and assumptions (guesses) may be necessary, yet a good approximate answer can generally be found, though perhaps not with ease or simplicity!

Here is a rather elaborate example of such a calculation.

In a chemically defined medium with glucose as sole source of carbon the organism *Bacillus megaterium* grows at 37 °C with a doubling time of 90 min. In an extract of these bacteria the enzyme pyruvate carboxylase has an activity of only 30 nmol min^{-1} (mg protein)$^{-1}$. This enzyme is the sole means by which a net synthesis of oxaloacetate (and hence of products derived from oxaloacetate) occurs in these organisms. Can the observed enzymic activity account for the rate of growth?

First calculate what is the rate (per minute) at which new organisms are made per mg of existing bacteria, that is, the specific rate of growth (μ):

$$\mu = 0.693/t_d (\text{see Chapter 5, about logarithms})$$
$$\therefore \mu = (0.693/90)\,min^{-1} = 0.0077\,min^{-1}$$

So, 1 mg of existing bacteria makes 0.0077 mg new bacteria per minute.

Now, suppose that protein represents half of the total dry weight of the bacteria. Hence, we can say that 0.5 mg of existing protein makes 0.0077 mg new bacteria per minute.

Thus, 1 mg existing protein makes 15.4 µg new bacteria min^{-1}.

Next, estimate the quantity of oxaloacetate that must be made to allow 1 mg of new bacteria to be assembled. This is where some approximations and knowledge of metabolic pathways are needed, and we decide that a total of about 250 µg oxaloacetate is needed. Or 250 / 132 µmoles oxaloacetate (= 1.89 µmoles) are needed to make 1 mg bacteria.

Finally, put the calculations together:

If 1890 nmoles oxaloacetate are needed to make 1 mg bacteria

Then 1.890 nmoles oxaloacetate are needed to make 1 μg bacteria

So 1.89×15.4 nmoles ($= 29$ nmoles) are needed to make 15.4 μg new
bacteria

From all this we conclude that oxaloacetate must be made by pyruvate
carboxylase at a rate ≥ 29 nmol $\min^{-1}$(mg protein)$^{-1}$ to be consistent with a
doubling time of 90 min. Thus, the observed activity of this enzyme is
sufficient, though with not much to spare!

The activity of pyruvate carboxylase in the intact organisms cannot be
much lower than we have calculated (or else growth would have to be
slower), but the activity might be higher in the undamaged organisms
than in an extract. If the activity in the extract had been considerably higher
than appears (from the calculation) to be necessary for the rate of growth,
then one would have reason to wonder whether the intracellular enzyme
was really not so active as the extracted enzyme, perhaps because of an
unfavourable internal pH value, or a limited supply of the substrates,
pyruvate or CO_2, within the organisms.

A variation of the calculation that we have just shown is to determine
the maximum rate of growth (t_d or μ) that could be achieved when some
vital enzyme has a given specific activity. The steps in the new calculation
are essentially those already considered, but done in reverse order. Here is
an example:

The enzymes glutamine synthetase (i) and glutamate synthase (ii)
together may allow a net assimilation of ammonia (as glutamate) by bacteria
growing in minimal medium with low concentrations of ammonia:

(i) Glutamate + NH_3 + ATP $\rightarrow$ Glutamine + ADP + P_i

(ii) Glutamine + α-oxoglutarate + NADPH + H^+ $\rightarrow$ 2 Glutamate +
NADP$^+$

(1) *Escherichia coli* strain w was grown with limiting ammonia ($t_d = 50$ min)
and an extract was then made. In the assay for glutamine synthetase the
initial rate of glutamate-dependent release of inorganic phosphate was
370 nmoles $\min^{-1}$ with 0.1 ml of undiluted extract. In the assay for
glutamate synthase the initial rate of glutamine-dependent oxidation of
NADPH was 180 nmoles $\min^{-1}$ with 50 μl of the same extract. A sample

(0.3 ml) of the extract after dilution (1 ml + 99 ml) contained 27 µg of protein.

Could these two enzymes alone be responsible for assimilation of ammonia by these organisms? Calculate what would need to be the minimum specific activity of each enzyme (µmoles min^{-1} (mg protein)$^{-1}$) to be consistent with the observed doubling time. Compare these estimates with the measured specific activities of the enzymes. Assume that 1 g (dry wt) of *E. coli* organisms contains 0.5 g of protein and that 0.15 g of nitrogen must be assimilated to synthesise 1 g (dry wt) of new organisms. (The specific growth rate is given by 0.693 / t_d, and the atomic wt of nitrogen is 14 Da.)

(2) *Escherichia coli* w was next grown with excess ammonia (t_d = 30 min) and other conditions unaltered, then an extract was made. In the assay for glutamine synthetase the initial rate of glutamate-dependent release of inorganic phosphate was 210 nmoles min^{-1} with 0.4 ml of this undiluted extract, and in the assay for glutamate synthase the initial rate of glutamine-dependent oxidation of NADPH was 260 nmoles min^{-1} with 50 µl of the same extract. A sample (0.25 ml) of this extract after dilution (1 ml + 99 ml) contained 30 µg of protein.

After making the same assumptions as before, what can you deduce about assimilation of ammonia when it is available in excess? Suggest further enzymic assays that might test your hypothesis.

Answer

(1) Growth with limiting ammonia

Glutamine synthetase activity = 370 nmoles min^{-1} with 0.1 ml extract

$\equiv$ 3700 nmoles min^{-1} ml^{-1}

27 µg protein in 0.3 ml extract (dil. 1 + 99)

$\equiv$ (27 ÷ 0.3) × 100 µg ml^{-1} ∴ 9 mg ml^{-1} undil.

∴ Specific activity = 3700 ÷ 9 = **411 nmoles min^{-1} (mg protein)$^{-1}$**

Glutamate synthase activity = 180 nmoles min^{-1} with 0.05 ml extract

$\equiv$ 3600 nmoles min^{-1} ml^{-1}

∴ Specific activity = 3600 ÷ 9 = **400 nmoles min^{-1} (mg protein)$^{-1}$**

$t_d = 50$ min $\therefore$ specific growth rate $= 0.693 \div 50 = 0.013\,86$ min^{-1}

i.e. 0.013 86 g new organisms min^{-1} made by 1 g existing organisms

$\equiv 13.86$ µg new organisms min^{-1} made by 1 mg existing organisms (containing 0.5 mg protein)

Hence, 27.72 µg new organisms min^{-1} can be made by 1 mg protein

0.15 µg nitrogen needed to make 1 µg organisms

Hence, 27.72×0.15 µg nitrogen must be assimilated in 1 min by 1 mg protein

$$\equiv (27.72 \times 0.15) \div 14 \text{ µg atoms of nitrogen (or µmoles of ammonia)}$$

$$\equiv 0.297 \text{ µmoles ammonia must be assimilated min}^{-1} \text{ (mg protein)}^{-1}$$

Thus, both enzymes must have specific activities of approx. 300 nmoles min^{-1} (mg protein)$^{-1}$ to be consistent with the observed rate of growth. Since both activities are higher than this, then the two enzymes alone **could** be responsible for the assimilation of ammonia.

(2) Growth with excess ammonia

    Glutamine synthetase activity $= 210$ nmoles min^{-1} with 0.4 ml extract $\therefore$ 525 nmoles min^{-1} ml^{-1}

    30 µg protein in 0.25 ml extract (dil. 1 + 99)

    $\therefore 30 \div 0.25 \times 100$ µg ml$^{-1} \equiv 12$ mg ml^{-1} undil.

    $\therefore$ Specific activity $= 525 \div 12 = $ **44 nmoles min^{-1} (mg protein)$^{-1}$**

The doubling time is now 30 min and so the necessary rate of assimilation of ammonia is $297 \times 50/30 = 495$ nmoles min^{-1} (mg protein)$^{-1}$. Because the two enzymes act sequentially, the overall rate of assimilation by these two enzymes cannot be greater than that of glutamine synthetase (above) which is far too low. Therefore, when ammonia is present in excess it must be assimilated by some other process.

(The specific activity of glutamate synthase in this extract is 433 nmoles min^{-1} (mg protein)$^{-1}$, but it is not strictly necessary to work this out.)

This other process may be the action of the enzyme **glutamate dehydrogenase**:

$$\alpha\text{-oxoglutarate} + NH_3 + NADPH + H^+ \rightarrow \text{glutamate} + H_2O + NADP^+$$

which has a high K_M value for ammonia, but which does not consume ATP in synthesising glutamate. The two extracts should be assayed for this enzyme, and it might be expected to be found with high activity (>500 nmoles min^{-1} (mg protein)$^{-1}$) in the extract of organisms grown with excess ammonia. It may or may not be present in the other extract.

9 | Spectrophotometry

And when we consider that other theory of the natural philosophers, that all other earthly hues – every stately or lovely emblazoning – the sweet tinges of skies and woods; yea, and the gilded velvets of butterflies, and the butterfly cheeks of young girls; all these are but subtle deceits, not actually inherent in substances, but only laid on from without; so that all deified Nature absolutely paints like the harlot, whose allurements cover nothing but the charnel-house within; and when we proceed further, and consider that the mystical cosmetic which produces every one of her hues, the great principle of light, for ever remains white or colorless in itself, and if operating without medium upon matter, would touch all objects, even tulips and roses, with its own blank tinge –

Herman Melville

Three commonplace observations are the basis of this subject:

(1) A stronger solution of a coloured substance looks darker than does a weaker solution of that substance.
(2) A thin layer of a coloured solution is paler than a thicker layer of the same solution.
(3) Different substances can give solutions of different colours.

9.1 Use of the spectrophotometer

These observations are made quantitative by using a **spectrophotometer**. In this machine a beam of light from an electric bulb (wavelengths about 350 to 800 nm) or from a source of UV (wavelengths about 220 to 350 nm) passes through a solution. The wavelength of this light can be precisely defined (any integer value in the range between about 220 and 800 nm may be chosen) and the solution is held in a cuvette made of a material (silica for UV wavelengths, glass or plastic for visible light) that is transparent to light of the wavelength being used. The pathlength of the beam of light through the solution is almost always 1 cm. The cuvette, depending on its size, may hold maximally 1 ml or 3 ml of solution, or may contain slightly smaller volumes of liquid than these, provided that the entire beam of light still can pass through the solution below the liquid meniscus.

The simpler kind of apparatus measures the intensity of the beam of light when it goes through a liquid (usually water or buffer) that absorbs very little light at the wavelength employed. This blank is used to set the scale of the spectrophotometer to zero extinction or 100% transmittance, both of which lie at the same place on the scale (see below). Then the cuvette with the test solution (in the same solvent as the blank) is moved into the beam, and the intensity of the transmitted light is measured. The scale usually shows both:

the extinction (E) of the test solution, that is: $\log_{10}(I_0/I)$, and the % transmittance (T), $100\ I/I_0$

where I_0 and I are the intensities of the light transmitted by the blank and by the test solutions respectively. (If $E = 0$ then it follows that $I_0\ /\ I = 1$. Hence $I\ /\ I_0 = 1$ and $T = 100$.)

E and T can be related to each other:

$$E = 2 - \log_{10} T \qquad\qquad T = 10^{(2-E)}$$

(See if you can do this fairly easy interconversion.)

Single-beam spectrophotometers are widely used for colorimetric and turbidimetric assays.

A source of light, such as an electric light bulb, does not emit light with equal intensity at all wavelengths, and the photoelectric cell, which in the spectrophotometer measures the intensity of the transmitted light, is also not equally responsive to light of all wavelengths. For these reasons, the I_0 setting must be established for each separate wavelength that is used. As well, a solution that is the blank in a colorimetric assay (see below) may absorb some light and so give an I_0 value that is lower than pure water would give at the same wavelength.

By experiment one finds that the relation (at a given wavelength of light) between the intensity of the light transmitted by a solution (I) and the concentration of a coloured solute (C) is not linear (Fig. 9.1).

Increasing C has a progressively smaller effect in lowering the value of I, or in other words, the smaller I becomes, the more slowly it decreases as C

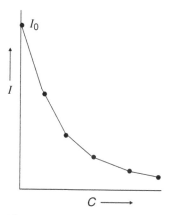

Fig. 9.1 The relation between the intensity of the light transmitted by a solution (I) and the concentration of a coloured solute (C); I_0 is the intensity of the transmitted light when C is zero.

continues to increase. This parallels the relation between the disintegrations per minute of a radioactive substance and the time at which counts are measured (see Chapter 11), and it follows that we can write:

Beer's law $I = I_0 e^{-ac}$

when the lightpath is 1 cm throughout; a is a constant.

I decreases in the same way as the pathlength (L cm) is increased, and so we can also write:

Lambert's law $I = I_0 e^{-\alpha L}$

when the concentration is the same throughout; α is a constant.

Why does I decrease exponentially rather than linearly with C (and with L)? Although it is true that low values of I cannot be measured accurately this is not the reason for the non-linearity. The correct interpretation is shown in Fig. 9.2.

We saw that Beer's law states that $I = I_0 e^{-ac}$ and Lambert's law states that $I = I_0 e^{-\alpha L}$. These two equations can be combined to give **the Beer–Lambert law**:

$$I = I_0 e^{-acL}$$

The Beer–Lambert law can be rearranged:

$$I/I_0 = 1/e^{acL}, \text{ and so:}$$
$$e^{acL} = I_0/I$$

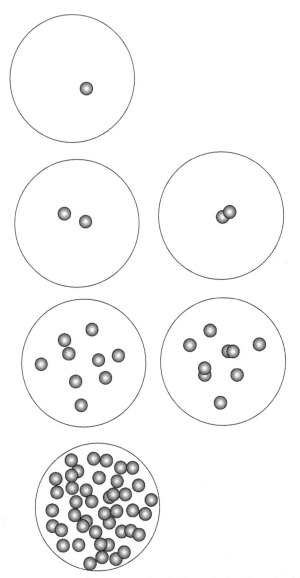

Fig. 9.2 Light-absorbing molecules in a solution through which a circular beam of light is passed. Two molecules will not absorb quite twice as much light as one, because Brownian motion will cause one molecule occasionally to move behind the other and be hidden. As more and more molecules are present the addition of another will have less and less effect of increasing the light absorbed. Only rarely will all the molecules absorb maximally.

from which it follows that $\ln (I_0 / I) = \mathbf{ac}L$, and hence

$$\log_{10}(I_0/I) = acL/2.303 \equiv (a/2.303)cL$$

Here, a is a constant, and so a / 2.303 is also a constant.

When a / 2.303 is replaced by A and $\log_{10}$ (I_0 / I) is replaced by E we have

$E = AcL$ (the form in which this law is usually learned)

where:

> E = extinction of the solution at a given wavelength of light (a logarithm and so a pure number)
>
> A = molar extinction coefficient of the solute at the given wavelength (litres mole^{-1} cm^{-1})
>
> L = thickness of the solution (= lightpath of cuvette) (cm)

The equation for this law can be rearranged in several ways:

(1) $A = E/cL$

which explains the rather strange dimensions of A. (The molar extinction coefficient is numerically equal to the extinction of a 1 M solution with a 1-cm lightpath.)

(2) $E = ALc$

which shows that the extinction rises as concentration and lightpath increase.

(3) $c = E/AL$

If, as is virtually always the case, $L = 1$ cm, then **the Beer–Lambert law predicts a linear relation between the extinction of a solution at a given wavelength and the concentration of the solute** that absorbs the light. The relationship departs from linearity at high concentrations of solute, when too little light emerges from the solution for I to be measured accurately, and the value of E does not then increase as much as the equation predicts for a given rise in c.

9.2   Colorimetric assays

Many coloured substances can be measured in solution by determining E. If A and L are known, then c may be calculated at once. Often though, A is not known. Instead, we measure the E values given by known concentrations (or amounts, see below) of the substance (S) being assayed. Plotting these E values against concentrations (or amounts, see below) of S leads to a standard curve. The E value(s) of the test solution of S can then be read from the curve as a concentration or amount of S. The procedure is illustrated for the Folin–Ciocalteau assay that is discussed below.

A colourless substance may also be assayed if it can be converted quantitatively to a coloured derivative. This idea is the basis of very many procedures. Amino acids give a colour after reaction with ninhydrin; many sugars break down in hot, strong sulphuric acid to give a product that yields a colour when a suitable coupler, such as phenol or anthrone, is added; inorganic phosphate gives a blue colour in the presence of molybdate and a reducing agent. Some of these methods are quite elaborate, and difficult to do successfully without practice. Reagents have to be made up carefully and the working instructions must be followed closely. Even so, the standard curve may not be precisely reproducible, and it is best always to measure some standards (known amounts of the test material) together with the unknowns. Note that what is measured is a coloured derivative of the substance being assayed. The chemical structure of such a derivative may not be known, and so a molar extinction coefficient cannot always be determined.

Concentrations versus amounts in quantitative spectrophotometric assays

In these assays we generally set up a series of tubes, each one containing the same volume of the assay mixture, and to these we add increasing amounts of the authentic standard. Where necessary, water is also added, so that every tube in the series finally contains the same volume of liquid. It follows, therefore, that the concentration of the standard in any tube will be directly proportional to the amount of the standard added. Hence, **the extinction developed in the assay will be directly proportional to the amount of standard added**. The samples of the test solution are also made up to the same total volume in the assay. Hence, **the extinctions of the test samples in the assay will also be proportional to the amount of test substance present**.

Many people are so wedded to the idea that extinction is a function of concentration (which of course it is) that they find it difficult to understand that extinction can also be a function of amount (which it may) when the volume remains constant. In plotting a standard curve for an assay it is perfectly valid practice to label the x axis as amount of standard (e.g. protein (μg); glucose (μmoles)) rather than as concentration of standard. Not only is this valid, but it makes subsequent calculations very much easier. An example will be considered at length to show the advantages.

In the **Folin–Ciocalteau assay for protein** one takes 0.5 ml of the standard or the test solutions (water is added when necessary to make up to this fixed volume), then reagent A (2 ml) and reagent B (0.2 ml) are added, so that the total volume of liquid is 2.7 ml in every tube. The extinctions of the mixtures (which become blue when protein is present) are read at 750 nm in a cuvette of 1-cm lightpath.

Tube no.	1	2	3	4	5	6	7
Standard (200 μg protein ml^{-1}) (ml)	0	0.1	0.2	0.3	0.5	0	0
Water (ml)		0.5	0.4	0.3	0.2	0	0.4	0
Test solution (ml)		0	0	0	0	0	0.1	0.5
Amount of protein in tube (μg)		0	20	40	60	100	?	?
Extinction at 750 nm		0.00	0.18	0.35	0.51	0.86	0.13	0.59

Plot **amount of protein (μg)** on the x axis, against **extinctions**, measured against tube 1 as zero.

One then can read from the graph that:

- 14 μg of protein was present in tube 6. This protein must have been present in 0.1 ml of test solution, so that the solution contained 140 μg protein ml^{-1}.

- 70 μg of protein was present in tube 7. This protein must have been present in 0.5 ml of test solution, so that the solution contained 140 μg protein ml^{-1}.

Estimates of the protein content of the test solution may be averaged if all are felt to be equally reliable. (Sometimes one or more of the extinctions given by different volumes of the test solution falls outside the range covered by the standards. In such cases these results ought not to be included in an average.)

One could reasonably plot volume of standard solution (ml) on the x axis against extinction. This is still plotting amounts, but the subsequent calculations are a little more complicated and can lead to confusion. The extinction of a test solution would be read off as the volume of standard that gave the same extinction, and it would be necessary then to find how much protein was present in this volume of standard. The test solution would have contained that same amount of protein if the volume of test solution had been assayed. It is all too easy to get mixed up between ml of standard and ml of test solution.

If, in spite of everything, one still wishes correctly to plot concentration against extinction, then the calculations really become intricate. Realise that the actual concentrations of protein in the standards (when the extinctions are measured) are 20 µg / 2.7 ml ($\equiv$ 7.4 µg ml^{-1}); 40 µg / 2.7 ml ($\equiv$ 14.8 µg ml^{-1}) etc. These awkward numbers could be plotted, but they certainly do not make life easy in subsequent calculations.

9.3   Optical density

Turbid suspensions: optical density and extinction

Some of the light is scattered when it passes through a turbid suspension, so that the value of I_0 will be greater than I. A spectrophotometer will measure $\log_{10}(I_0 / I)$. When this function is due to scattering of light, rather than absorption, it is called optical density and not extinction. The optical density does not precisely obey the Beer–Lambert law because scattering is due to the size, shape and refractive index of the suspended particles rather than to their molecular weight and absorbency. Since the particles are not in solution, their concentration cannot be expressed in terms of molarity, but only as weight or number per unit volume.

Measurement of optical density is a quick and convenient way of assessing growth of many microorganisms, because, within limits, the optical density is directly proportional to the number of organisms in a suspension. Fewer than about 1×10^7 organisms ml^{-1} give no perceptible turbidity and more than about 5×10^9 organisms ml^{-1} are too turbid for accurate measurement, though a dense suspension can, of course, be diluted and then measured.

Microorganisms of different species may have different sizes or shapes; even the organisms from a pure culture may change in these respects when grown under altered conditions. Hence, the relation between optical density and the concentration of a suspension of organisms can also be quite variable, and needs to be found by experiment.

A dense suspension of the organisms is grown under appropriate conditions, diluted and counted microscopically. As well, organisms from a known volume of the suspension can be collected, washed and dried and their weight determined. Various dilutions of the suspension are made, with known numbers or known weight of organisms ml^{-1}, and the optical densities of these suspensions are measured. The resulting data allow calibration curves to be drawn, which relate number of organisms ml^{-1} (x axis) or dry weight of organisms ml^{-1} to optical density (y axis).

Optical density plotted against time: microbial growth curves

Growth of a culture is often followed by reading its optical density at intervals. Mistakes in the interpretation of the results can occur by failing to realise that:

(1) It is almost certain that the organisms will be in the exponential phase of growth before turbidity becomes detectable (unless a very large inoculum had been used, so that turbidity could be seen immediately after inoculation).

(2) Optical density is not an exponential function (it is logarithmic), and is directly proportional to the number of organisms ml^{-1}. This means that **the logarithm of the optical density must be plotted** (even though the optical density itself is defined as a logarithm) against time in order to show the duration of the exponential phase of growth and to evaluate the doubling time during this phase. Plotting the optical density itself is equivalent to plotting number of organisms ml^{-1} rather than the logarithm of the number of organisms ml^{-1}.

9.4   Absorption spectra

Pure water is colourless because it is transparent to visible light. A solution looks black when no light can pass through it. A coloured solution absorbs light of some wavelengths but is transparent to light of other wavelengths. For example, a blue solution appears so because it is absorbing light in the red–yellow region of the visible spectrum, and is allowing the shorter wavelength blue light to pass. Conversely, a solution may look orange

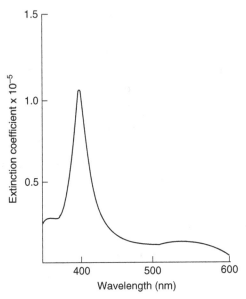

Fig. 9.3 Absorption spectrum of the oxidised form of cytochrome *c*.

because it is absorbing light in the blue region. Solutions which absorb only UV light will appear colourless to the eye. Which wavelengths are absorbed, and which are not, is displayed graphically by the absorption spectrum of a coloured solution.

The **absorption spectrum** must be determined experimentally. Light of a single wavelength is directed at the solution, so that the extinction can be established at that wavelength. This process is repeated at a succession of single wavelengths until a graph can be drawn (Fig. 9.3) of extinction versus wavelength. As well as recording the wavelength at which E is at a maximum it is valuable (particularly with purines and pyrimidines) to note the ratio of E at 280 nm to E at 260 nm. In addition the pH value of the solution should be known because the shape of the absorption spectrum is sensitive to changes in acidity.

When we need to measure the absorbance of a coloured solution in order to find its concentration, it is usually best to choose the wavelength at which the absorbency is highest, so as to make the sensitivity greatest. The value of A (the molar extinction coefficient) will be highest at the wavelength of maximum absorption, and will be different (i.e. lower) at other wavelengths. Sometimes a solution which has an extinction that is too high (at the absorption maximum) for accurate measurement can be assayed successfully by reading the extinction at a different wavelength, where the extinction is less. In such a case it will be necessary to know the molar extinction coefficient (or to measure the standard curve) at this other wavelength.

The absorption spectrum is also sometimes helpful in identifying a substance. For instance, the spectra of solutions of the two purines and the two pyrimidines from DNA differ from each other in the UV region, and each spectrum alters characteristically as the pH value of the aqueous solvent is changed. Nevertheless, absorption spectra are much less useful in identification of chemical substances than are infrared or nuclear magnetic resonance spectra.

Often one encounters a coloured solution that has an extinction too great for accurate measurement. Remember that if the extinction is 2 then only 1% of the incident light is getting through the cuvette. Does diluting the solution (until the extinction becomes a valve that can be read) allow the concentration of the undiluted solution to be calculated reliably? The answer is 'yes sometimes', but there are many pitfalls.

If the solution is of a single coloured solute in water, then dilution will be safe. The same is true for a turbid suspension of organisms in water. However, in many spectrophotometric assays we are dealing with a complex solution, which may contain more of other liquids (e.g. ethanol, acetic acid, sulphuric acid) than of water. The colour developed in the assay might only be stable in the particular solution that was used. In such cases dilution with water may lead to wrong results – the colour is likely to be decreased by more than the extent of dilution would predict, or the solution may become turbid. The only really safe procedure is to repeat the assay with a smaller amount of the test material so that an extinction can be found which falls within the range of the standard curve.

The shape of standard curves

A graph of extinction (y axis) against concentration (x axis) should be linear if Beer's law is obeyed. Virtually always, though, the curve becomes flattened as E rises above some limit. There can be several reasons for this, for instance: when I becomes small it cannot be measured accurately; the yield of coloured product in an assay may not be proportional to c at higher values of c. Readings from the non-linear part of the curve must be unreliable because a small change in E may represent a large change in c.

9.5   Following the course of an enzymic reaction

In a **double-beam spectrophotometer** the contents of two cuvettes can be examined simultaneously. One beam of light passes through the blank cuvette and another beam (of the same intensity and wavelength) passes through the test cuvette while the intensities of the two transmitted beams are measured continuously. A recorder can then trace, for example, a change in the extinction of the test solution, relative to the blank, over a period of time. This kind of apparatus is particularly valuable for monitoring the course of an enzymic reaction, especially as the holder of the cuvettes can usually be kept at a suitable temperature, such as 30 °C.

For example, the progress of an oxidation reaction that leads to the reduction of NAD^+ to NADH can be followed by recording the increase of E at 340 nm. This is possible because NADH has a much higher molar extinction coefficient (6.22×10^3) at this wavelength than does NAD^+ (see Fig. 3.1).

10 | Energy metabolism

*The First Law of Thermodynamics: The total amount of energy in Nature is constant.
The Second Law of Thermodynamics: The total amount of entropy in Nature is
increasing.*

Living organisms, even the most simple, are extremely improbable struc-
tures. Their highly complex forms are not at all likely to appear by the
random associations of small molecules (spontaneous generation); and the
elaborate organic compounds that make up an organism are thermody-
namically unstable – they tend to break down with a release of energy and
an increase of disorder (entropy). Energy must be available to drive forward
the reactions of biosynthesis, that is to move from disorder to order, and to
maintain the state of order against a hostile Nature. Energy may be needed
for other purposes too, such as movement, uptake of nutrients, or maintenance of body temperature by some animals.

10.1   Sources of energy

Energy is gained by living organisms from two main sources:

(1) **Light** (i.e. by photosynthesis). Photosynthesis may be considered as the
    fundamental process by which most living organisms gain energy.
    However, it is easier to approach the complicated ideas within the
    subject of energy metabolism by first considering oxidation reactions.
(2) **Oxidation reactions** (usually of an organic molecule, and usually with
    oxygen as the final electron acceptor).

Many organic molecules can be broken down by microbes (and other
living organisms) in the presence of oxygen, to yield CO_2 and water and to
release energy. This energy is trapped initially as ATP, which is used to drive
many cellular processes.

The decomposition of an organic substrate takes place as a sequence of reactions, in which molecules of a cofactor are reduced by various compounds derived from the initial substrate. For example, consider the complete oxidation of methanol (CH_3OH) to CO_2 and water by NAD^+ and O_2 with appropriate enzymes:

$$CH_3OH + NAD^+ \rightarrow H_2CO \text{ (formaldehyde)} + NADH + H^+$$
$$H_2CO + NAD^+ + H_2O \rightarrow HCOO^- \text{(formate)} + H^+ + NADH + H^+$$
$$HCOO^- + H^+ + NAD^+ \rightarrow CO_2 + NADH + H^+$$

Overall, so far: $CH_3OH + 3NAD^+ + H_2O \rightarrow CO_2 + 3NADH + 3H^+$

The molecules of reduced cofactor are re-oxidised by passage of electrons and protons through a sequence of carrier molecules, the electron-transport chain, to the final electron acceptor which (in most cases) is molecular oxygen from the air. As a result of this transfer process ADP is phosphorylated to give ATP (2ATP formed per $NADH + H^+$ oxidised):

$$3NADH + 3H^+ + 1\frac{1}{2}O_2 \rightarrow 3NAD^+ + 3H_2O$$
$$6ADP + 6P_i \rightarrow 6ATP + 6H_2O$$

Putting the equations together we get overall:

$$CH_3OH + 6ADP + 6P_i + 1\frac{1}{2}O_2 \rightarrow CO_2 + 8H_2O + 6ATP,$$

a correct summation which conceals the real complexity of the process.

Most non-photosynthetic forms of life can generate reduced cofactor from only organic electron-donors. However, some kinds of bacteria can use inorganic molecules to reduce cofactors. For example, $H_2 + NAD^+ \rightarrow NADH + H^+$. These organisms are generally able to use the reducing power (i.e. NADH) and the ATP derived from the oxidation of the inorganic substrate to make all their cellular components from CO_2 as sole source of carbon, and are therefore called **autotrophs**, in particular named as **chemolithotrophs**.

In principle, any non-toxic inorganic reducing agent of sufficient reducing potential (see later) might be used to form a reduced cofactor, from which energy could be gained by its re-oxidation. In practice, only a few inorganic materials that occur widely in Nature are used in this way by microorganisms. The main substrates are:

Hydrogen gas (H_2) – oxidised to water

Thiosulphate ($S_2O_3^{2-}$) and other reduced inorganic forms of sulphur (S^0, S^{2-} etc.) – oxidised to sulphate

Ammonium ions (NH_4^+) – oxidised to nitrate

Nitrite ions (NO_2^-) – oxidised to nitrate

Ferrous ions (Fe^{2+}) – oxidised to Fe^{3+}

Carbon monoxide (CO) – oxidised to CO_2

Usually, a given species cannot employ many of these electron-donors. Thus, hydrogen bacteria do not oxidise thiosulphate, and oxidisers of ammonium ions cannot use ferrous ions.

Some of these inorganic donors are not sufficiently powerful reducing agents to be able to donate electrons to NAD^+. They can only reduce cofactors that are better oxidising agents (e.g. FAD or cytochromes) with more positive electrode potentials than NAD^+.

Electrode potentials

In an oxidation reaction the substance being oxidised loses electrons (i.e. is itself an electron-donor or reducing agent), and the oxidising agent is the electron-acceptor. Hence, an oxidation process may be split, formally, into two steps, as for example:

(1) A donating half-reaction: $H_2 \rightarrow 2H^+ + 2e^-$
(2) An accepting half-reaction: $NAD^+ + 2H^+ + 2e^- \rightarrow NADH + H^+$.

The strength of a half-reaction as a reducing agent is expressed as volts and written as E_0' (at pH 7). The more negative is this potential, the more powerful is the half-reaction as an electron-donor (reducing agent). The more positive is the potential, the more powerful the half-reaction as an electron-acceptor (oxidising agent).

Note that the sign and the numerical value of E_0' do not change even when the half-reaction is written in the reverse direction:

$$H_2O \rightarrow \tfrac{1}{2}O_2 + 2H^+ + 2e^- \quad \text{and} \quad \tfrac{1}{2}O_2 + 2H^+ + 2e^- \rightarrow H_2O$$

both having the same E_0' value (+0.82 V).

However, the actual value of a half-reaction (E') is dependent on pH value, and even at a given pH value is not constant, but depends on the

Table 10.1 *Some electrode potentials at pH 7*

	E_0' (V)
Ferredoxin$_{(red)}$ → Ferredoxin$_{(ox)}$ + e$^-$	−0.43
NADH + H$^+$ → NAD$^+$ + 2H$^+$ + 2e$^-$	−0.32
FADH$_2$ → FAD + 2H$^+$ + 2e$^-$	−0.03
Cytochrome $c_{(red)}$ → Cytochrome $c_{(ox)}$ + e$^-$	+0.22
Cytochrome $a_{(red)}$ → Cytochrome $a_{(ox)}$ + e$^-$	+0.50
H$_2$O → ½O$_2$ + 2H$^+$ + 2e$^-$	+0.82
CO + H$_2$O → CO$_2$ + 2H$^+$ + 2e$^-$	−0.53
H$_2$ → 2H$^+$ + 2e$^-$	−0.42
S$_2$O$_3^{2-}$ + 5H$_2$O → 2SO$_4^{2-}$ + 10H$^+$ + 8e$^-$	−0.25
NH$_4^+$ + 2H$_2$O → NO$_2^-$ + 8H$^+$ + 6e$^-$	+0.35
Fe^{2+} + (H$^+$) → Fe^{3+} + e$^-$ + (H$^+$)	+0.77

molar concentration of the electron-donor and on the relative concentrations of its oxidised (c_{ox}) and reduced (c_{red}) components:

$$E' = E_0' + (0.06/n) \times \log_{10}(c_{ox}/c_{red}) \text{ at } 30°C$$

where n is the number of electrons on one side of the half-reaction. Consequently, the potential becomes more negative as the ratio c_{ox} / c_{red} diminishes, and is equal to E_0' only when c_{ox} / c_{red} has a value of 1.

Electrons from a half-reaction that has the more negative potential will be accepted by any half-reaction that is less negative, so that the acceptor reaction is driven from right (as written in Table 10.1) to left. For example, electrons from NADH will reduce molecular oxygen to water:

$$NADH + H^+ \rightarrow NAD^+ + 2H^+ + 2e^-$$
$${}^1\!/_2O_2 + 2H^+ + 2e^- \rightarrow H_2O$$

10.2   Determining the free energy change in an oxidation–reduction reaction

To do this, subtract the signed potential of the electron-donating half-reaction from the signed potential of the electron-accepting half-reaction to give $\Delta E_0'$ and determine the number of electrons (n) transferred in the overall reaction. Hence, for the reaction between NADH + H$^+$ and ½O$_2$ we have $\Delta E_0'$ = +0.82 V − −0.32 V = +1.14 V, and n = 2.

Substitute these numbers into the formula: $\Delta G_0' = -n.\Delta E_0'. F$, where F is the Faraday constant (23 061 calories per V).

Hence, $\Delta G_0' = -2 \times 1.14 \times 23\,061 = -52\,579$ calories (-53 kcal) per mole of NADH oxidised by oxygen.

Bacteria are supposed to generate 2 moles of ATP (which synthesis requires about 20 kcal) from the oxidation of 1 mole of NADH. Thus, about 40% of the total free energy released by the oxidation is conserved as ATP.

Reverse electron flow

In order to grow with CO_2 as sole source of carbon, organisms need a reducing agent of enough power to drive the synthesis of cell substances:

$$CO_2 + 4H^+ + 4e^- \rightarrow CH_2O \text{ (cellular material)} + H_2O$$

This requires an electrode potential similar to that of NADH (about -0.3 V). Most of the inorganic substrates used for energy generation by autotrophs have potentials more positive than this. Electrons have to be driven, by spending energy, from a more positive to a less positive half-reaction. This process is called **reverse electron flow**. Much of the ATP produced by oxidation of the substrate is used in this way.

Use of terminal oxidants other than oxygen

Some bacteria can use agents other than oxygen to re-oxidise reduced cofactors. In such cases organic materials may be oxidised to CO_2 and water in the complete absence of air, and inorganic substrates can also be attacked. The process is called **anaerobic respiration**. The principal cases are shown in Table 10.2.

Table 10.2 *Electrode potentials of terminal oxidants at pH 7*

		E_0' (V)
$NO_3^- + 2H^+ + 2e^- \rightarrow NO_2^- + H_2O$	Nitrate reduction	$+0.51$
$NO_2^- + 2H^+ + e^- \rightarrow NO + H_2O$	Denitrification	$+0.36$
$SO_4^{2-} + 8H^+ + 8e^- \rightarrow S^{2-} + 4H_2O$	Sulphate reduction	-0.25
$CO_2 + 8H^+ + 8e^- \rightarrow CH_4 + 2H_2O$	Methane generation	-0.24
$(2H^+ + 2e^- \rightarrow H_2$	Partial oxidation)	-0.42
Fumarate $+ 2H^+ + 2e^- \rightarrow$ Succinate	Organic oxidant	$+0.03$

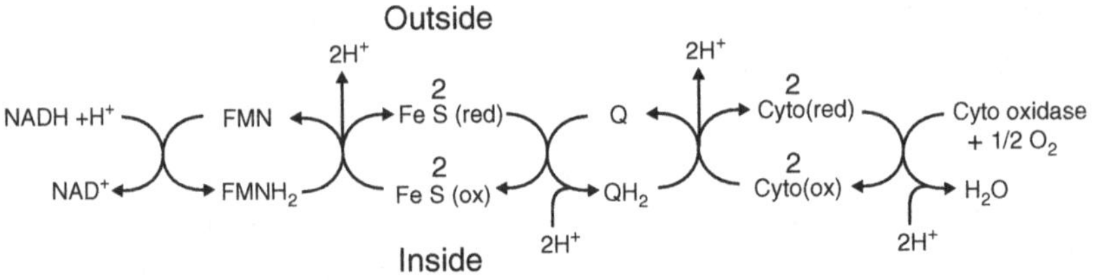

Fig. 10.1 An electron-transport chain. In this example four protons from the interior side of the membrane are consumed and four protons are transferred to the exterior side. The cofactors and enzymes are situated in the membrane.

The potentials of these half-reactions are less positive than is the reduction of oxygen to water. Hence, less energy is available when NADH is oxidised by NO_3^- (for instance 35 kcal per mole of NADH) than when O_2 is used (53 kcal per mole of NADH).

How is ATP formed during oxidation reactions?

Electrons from NADH do not pass directly to oxygen (or other final acceptor); instead the electrons travel through a sequence of carrier cofactors of increasingly positive E_0' values before reaching the final acceptor (see Fig. 10.1). These cofactors and the enzymes that catalyse the successive redox steps are located together in the cytoplasmic membrane of prokaryotes, and in the mitochondrial membrane of eukaryotes.

10.3 Proton motive force

The chemiosmotic theory proposes that the enzymes and cofactors of the electron-transport chain are positioned in the membrane in such a way that protons are expelled to the outside of the membrane by those reactions which release protons (Fig. 10.1). Protons are withdrawn from the cytoplasm (i.e. from the inside) by those reactions that cause an uptake of protons. In this way a gradient of concentration of protons and of electrical charge (higher and more positive on the outside) is established. The external pH value is 1.4 units lower (i.e more acidic) than inside, and the potential difference across the membrane is 0.14 V, with the outside being more

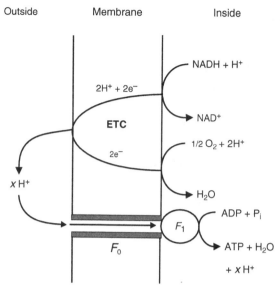

Fig. 10.2 Generation of ATP by the re-entry of protons. The electron-transport chain (ETC) carries protons to the exterior of the membrane. The return of these protons through the $F_0 + F_1$ complex drives the formation of ATP.

positive. The proton motive force Δp (volts) is made up of this membrane potential (E_m) and the chemical gradient ΔpH:

$$\Delta p = E_m - (2.3\ RT/F) \times \Delta pH$$
$$= E_m - 0.06 \times \Delta pH$$
$$= 0.14 - 0.06 \times -1.4 = 0.224\ V$$

From the equation $\Delta G_0' = -n.\Delta E_0'.F$ (see earlier) **the proton motive force corresponds to a free energy change of 5.2 kcal per gram ion of protons transported.**

This energy can be used to generate ATP when the proton motive force is dissipated by protons returning to the cytoplasm via a membrane-bound ATP synthase. The synthase is made up of two units: F_0, which spans the membrane and is the channel by which protons re-enter; and F_1, which is the enzyme that catalyses the formation of ATP (see Fig. 10.2). At least two protons must pass through the ATP synthase to generate one molecule of ATP from ADP + P_i because this reaction requires an input of 8 to 10 kcal per mole of ATP formed.

10.4   Fermentations

Very many microorganisms can grow on an organic substrate in the absence of oxygen or of any other electron-acceptors (such as inorganic materials or fumarate). Some of the substrate is partly degraded to provide energy in the form of ATP, and some is used to supply reduced carbon for biosynthesis, so that reduction of CO_2 is not a major process.

Because there is no terminal electron-acceptor, there is no generation of ATP by electron-transport phosphorylation. Instead, breakdown of the substrate leads to compounds that can phosphorylate ADP directly, a process called **substrate-level phosphorylation.**

For example, in the fermentation of glucose by the **Embden–Meyerhof–Parnas pathway** two molecules of 1,3-diphosphoglycerate and two molecules of phospho-enol-pyruvate are generated, and from these ATP is made:

2 1,3-diphosphoglycerate $+$ 2ADP $\rightarrow$ 2 3-phosphoglycerate $+$ 2ATP

2 phospho-enol-pyruvate $+$ 2ADP $\rightarrow$ 2 pyruvate $+$ 2ATP

(Although 4ATP molecules are generated by these reactions from one molecule of glucose, the net gain of ATP is only two molecules because two molecules of ATP have been consumed in producing fructose-1,6-bis-phosphate from glucose.)

During a fermentation one or more enzymic steps will usually produce NADH $+$ H$^+$, as for instance:

glyceraldehyde-3-phosphate $+$ P$_i$ $+$ NAD$^+$ $\rightarrow$ 1,3-diphosphoglycerate
$$+ \text{NADH} + \text{H}^+$$

The reduced cofactor must be re-oxidised by a compound that is an intermediate at a later step in the pathway, such as:

pyruvate $+$ NADH $+$ H$^+$ $\rightarrow$ lactate $+$ NAD$^+$

In this way overall redox balance is achieved without an external electron-acceptor. The average level of oxidation of the products of the fermentation is the same as that of the substrate:

$$C_6H_{12}O_6(\text{glucose}) + 2\text{ADP} + 2\text{P}_i \rightarrow 2C_3H_6O_3(\text{lactic acid})$$
$$+ 2\text{ATP} + 2H_2O$$

The substrate is only partly degraded in a fermentation, and so the amount of energy gained as ATP (2 molecules) per molecule of glucose used is much less than would be obtained if the substrate were completely oxidised to CO_2 and water (approximately 30 molecules of ATP per molecule of glucose).

Many organic compounds can be fermented by different microorganisms, though carbohydrates are generally the most widely used substrates. Investigations of the enzymic pathways and the chemistry of the products of fermentations have been a major area of classical biochemistry. One of the reasons for this interest is that some of the products of fermentations (such as ethyl alcohol, glycerol, butyric acid) are not without commercial importance.

10.5   Microbial photosynthesis

The basic feature of photosynthesis is the conversion of radiant energy (light) into chemical energy (ATP):

$$ADP + P_i + light = ATP + H_2O$$

This trapped energy may (or may not) then be used to drive the reduction of CO_2 to the level of cellular components ($C[H_2O]$).

$$CO_2 + 4H^+ + 4e^- (+ATP + H_2O)_n = C[H_2O] + H_2O(+ADP + P_i)_n$$

In this second reaction a reducing agent (source of $4H^+ + 4e^-$) is obviously needed.

Green plants and **cyanobacteria** gain this reductant by the photolysis of water:

$$H_2O + light = \frac{1}{2}O_2 + 2H^+ + 2e^- \ (E'_0 = +0.82)$$

These organisms contain chlorophyll a (absorption maximum 660 nm) as their main photosynthetic pigment. The primary event in photosynthesis is the expulsion of an electron from photosystem II (PS II):

$$PS\ II + light = PS\ II^+ + e^- \ (E'_0 = +1.0\ V)$$

Electrons released from PS II are raised by the energy of light to a negative potential such that they can be accepted by a series of electron-carrier molecules and passed to photosystem I (Fig. 10.3). A proton gradient is

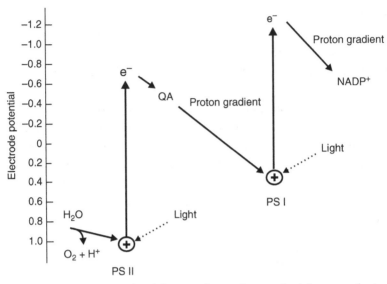

Fig. 10.3 Transfer of electrons during photosynthesis by green plants and cyanobacteria. PS I and PS II are photosystems I and II respectively.

created by this electron transfer, and in consequence ATP can be generated. Because of its strength as an oxidising agent, ions of PS II$^+$ can replace their lost electrons by removing electrons from water.

Photosystem I (PS I), in turn, when illuminated will expel electrons with enough energy to reduce NAD$^+$ to NADH + H$^+$. These lost electrons may be replaced from PS II (and hence from water) as described above, in which case the NADH + H$^+$ is available to reduce CO_2. Alternatively, the electrons may be returned to PS I through a series of carriers, with accompanying generation of a proton gradient, and hence formation of ATP. This latter process is called **cyclic photophosphorylation**, and does not lead to fixation of CO_2.

Other photosynthetic organisms

These are divided into four groups of bacteria: the purple sulphur, the purple non-sulphur, the green sulphur, and the green non-sulphur bacteria. None of these organisms contains chlorophyll *a*, and they all lack PS II so that they are unable to photolyse water, and do not, therefore, evolve oxygen during photosynthesis. The **sulphur bacteria** make up the majority of these

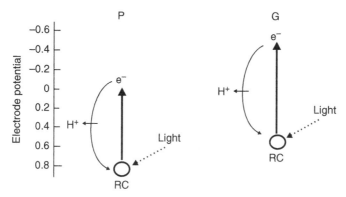

Fig. 10.4 Cyclic photophosphorylation in purple (P) and green (G) sulphur bacteria. RC is reaction centre. The arrow pointing down from the activated electrons to the reaction centres shows that the electrons return to the reaction centres, with the formation of a proton gradient (cyclic photophosphorylation).

photosynthetic species and are obligate anaerobes that use H_2S as reductant during fixation of CO_2:

$$CO_2 + 2H_2S = C[H_2O] + H_2O + 2S$$

The green sulphur bacteria contain bacteriochlorophyll *c*, with an absorption maximum at 740 nm, and the purple sulphur bacteria have bacteriochlorophylls *a* and *b* with maxima at 850 nm and 1020 nm respectively. Cyclic photophosphorylation by purple and green sulphur bacteria is illustrated in Fig. 10.4. In neither of these cases does the expulsion of an electron from the reaction centre leave a charged ion that has an electrode potential sufficiently positive to be able to remove electrons from water.

Electrons expelled by light are raised in the green sulphur bacteria to a negative potential that is enough to reduce $NADP^+$ to NADPH (Fig. 10.5). The electrons thus lost from the reaction centre are replaced by the oxidation of H_2S. The purple sulphur bacteria can also use H_2S to replace expelled electrons, but these expelled electrons do not gain a sufficient negative charge to reduce $NADP^+$ (Fig. 10.4). To achieve this reduction it is necessary for ATP to drive reversed electron flow.

The green and purple non-sulphur bacteria are usually **photoheterotrophs** which can grow aerobically in the dark with organic or inorganic sources of energy. Although they can use sulphur compounds as

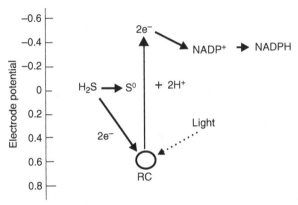

Fig. 10.5 Light-driven formation of NADPH by green sulphur bacteria. Electrons expelled from the reaction centre (RC) by light achieve a potential that is sufficiently negative to reduce $NADP^+$. Electrons lost from RC are replaced by oxidation of H_2S.

electron-donors for photosynthesis they are less dependent on these than are the sulphur bacteria.

Fixation of CO_2

Green plants and the cyanobacteria use the **Calvin** (or C3) **pathway** for converting CO_2 into biomass. The enzyme ribulose 1,5-bisphosphate carboxylase has a key role. This pathway is also used by most other photosynthetic bacteria and by aerobic autotrophs.

The reductive citric acid cycle occurs in green sulphur and some purple sulphur bacteria. This cycle requires an input of ATP and reduced ferredoxin and produces one molecule of acetyl coenzyme A from two molecules of CO_2.

Most of the purple sulphur bacteria lack some key enzymes of the citric acid cycle, and these organisms use the reductive acetyl coenzyme A pathway to fix CO_2. This is a linear route that reduces two molecules of CO_2 to acetyl coenzyme A by using H_2 and the enzyme hydrogenase, with an expenditure of ATP.

10.6   Photochemistry

Light is radiant energy of wavelengths between approximately 200 and 1000 nm. In some of its properties light behaves as a stream of particles called photons, each of which carries a fixed amount of energy that is directly related to the frequency (= speed of light in metres s^{-1} / wavelength in metres) of the light:

energy of one photon $= h\nu$

where h is Planck's constant (6.624×10^{-27} erg.s) and ν is the frequency (waves.s^{-1}). To produce a chemical change the light must be of a wavelength that is absorbed by the reacting substance.

The **law of photochemical equivalence** states that one photon causes a change in one absorbing molecule. Hence to bring about a change in one mole of absorbing molecules 6.023×10^{23} photons are needed. How much energy does this represent?

Chlorophyll a absorbs light at 660 nm, i.e. 660×10^{-9} m. The velocity of light is 2.9977×10^{8} m s^{-1} so that at a wavelength of 6.60×10^{-7} m light has a frequency of 2.9977×10^{8} / 6.60×10^{-7} s^{-1}.

6.023×10^{23} photons at 660 nm will therefore deliver:

$$6.023 \times 10^{23} \times 6.624 \times 10^{-27} \times 2.9977 \times 10^{8}/(6.60 \times 10^{-7})\ \textbf{ergs}$$

To convert from ergs to joules we divide by 1×10^{7} and then by 10^{3} to get kJ and finally by 4.184 to convert kJ into kcal (overall divide by 4.184×10^{10}) which means that we get a splendid equation:

$$6.023 \times 10^{23} \times 6.624 \times 10^{-27} \times 2.9977 \times 10^{8}$$
$$\div (6.60 \times 10^{-7} \times 4.184 \times 10^{10})\ \text{kcal}$$

on which to do a 'back of envelope' simplification.
Get rid of the some of the powers of 10:

$$6.023 \times 6.624 \times 2.9977 \times 10^{4}/(6.60 \times 4.184 \times 10^{3})$$

Do some rounding up and cancelling:

approx. $6 \times 1 \times 3 \times 10^{1}/(1 \times 4 \times 1) = 180/4$ **which is approx. 45 kcal**

So, we expect an answer about 40 when we use a calculator and avoid rounding. The precise answer rolls out as $= \textbf{43.3 kcal}$.

I hope that this looks straightforward. Really it isn't so easy. Getting the distance travelled by light (in 1 s) and the wavelength of the light to be in the same units (both in metres or both in centimetres) is vital, otherwise the answer can come out ruinously wrong!

If all of this energy were used by PS II to expel electrons, to what negative potential could the electrons be driven? The calculation is much simpler than the last one – take the equation:

$$\Delta G_0' = -n.\Delta E_0'.\ F$$

that we have already used to solve for $\Delta G_0'$ when we knew $\Delta E_0'$, and instead solve for $\Delta E_0'$, knowing that $\Delta G_0'$ is + 43.3 kcal. Therefore:

$$43\,300 = -1 \times \Delta E_0' \times 23\,061,\ \text{and}\ \Delta E_0' = 43\,300\ /-23\,061$$
$$= -1.88\,\text{V}$$

(Note that if F is expressed as 23 061 then $\Delta G_0'$ must be expressed in cal, not kcal.)

Since $\Delta E_0'$ is +1.0 V for the reaction PS II = PS II$^+$ + e$^-$ it follows that the electron expelled from PS II by a photon of light at 660 nm could be driven to approx. –0.9 V (see Fig. 10.3). Is **all** the energy of a photon really transferred to the expelled electron? The answer is very probably no, because some of the radiant energy is likely to be dissipated as heat.

In calculating the amount of energy delivered by light during a given time, such as 1 s, we need to know the intensity of the light as well as its frequency. The intensity is (in effect) the number of photons arriving per unit time per unit area. Light of 1 candela falling on an area of 1 cm^2 delivers approximately 2×10^{14} photons per second, the energy of these photons being dependent on the wavelength of the light. A dim source delivers relatively few photons in a given time, whereas a bright source delivers many more photons in the same time. Area is important too – the Sun delivers an enormous number of photons over a huge area, but what we need to know is how many photons arrive in the area occupied by the photosynthetic system, such as a leaf or a culture of microorganisms.

11 | Radioactivity

Time is nature's way to keep everything from happening at once.

John Wheeler

Every chemical element, except hydrogen (and then only isotope ^{1}H), has a **nucleus** made up of both **protons** and **neutrons**. The number of protons determines the chemical nature of the atom (e.g. hydrogen has one proton, helium two, carbon six, oxygen eight, and so on). The number of neutrons must, for reasons of stability, always be close to the number of protons in the nucleus, but is not necessarily the same. In the case of hydrogen, the one proton may be alone in the nucleus (^{1}H), or there may be one proton and one neutron (^{2}H, called deuterium), or one proton and two neutrons (^{3}H, called tritium). These different forms of hydrogen are called **isotopes**. Hydrogen and deuterium are both stable, but the tritium nucleus is unstable, and breaks down to helium by the emission of an electron, and so is radioactive:

$$^3\text{H} \rightarrow\, ^3\text{He} + e^-$$

The ^{3}He nucleus formed from the tritium contains two protons and one neutron, and is stable.

11.1 Isotopes

Isotopes of many other elements exist naturally, or can be produced industrially. Some are stable, some show a relatively slow radioactive decay, while some are very unstable, and decay extremely quickly. The rate of decay of a radioactive element may be expressed as its **half-life** (that is the time needed for half of the initial number of radioactive atoms to decompose) or by its decay constant (k; see Chapter 5), or by the average life of a radioactive atom ($1/k$), although this last usage has become very uncommon.

Table 11.1 *Some widely used radioisotopes and their half-lives*

^{3}H	12.26 years
^{14}C	5730 years
^{32}P	14.3 days
^{35}S	88 days
^{125}I	60 days

The different isotopes of a given element (e.g. ^{11}C, ^{12}C and ^{13}C) are chemically indistinguishable, and so an organic compound containing any one of these isotopes of carbon will react with an enzyme in the same way. This is the basis for the use of radioactive isotopes as tracers. A radioactive isotope can be incorporated into a molecule (e.g. generally labelled ^{14}C-glucose, in which all the carbon atoms are ^{14}C) which will be metabolised in exactly the same way as non-radioactive (^{12}C-) glucose. All the organic products that are derived from the carbon of the ^{14}C-glucose molecule will be radioactive, and this is an enormous help in recognising them.

The principal radioisotopes that are used in biological studies are ^{3}H, ^{14}C, ^{32}P and ^{35}S (Table 11.1). All of these isotopes emit electrons, while some other radioisotopes emit α-particles (protons) or γ- (gamma) rays. Oxygen and nitrogen do not have radioactive isotopes, although their stable 'heavy' isotopes ^{18}O and ^{15}N are often used as labels, but they have to be measured with a mass spectrometer, which is less convenient in most biological laboratories. Radioisotopes of many other elements (e.g. ^{59}Fe, ^{125}I) are frequently employed too.

Different radioisotopes emit electrons with different average amounts of energy, and the electrons from an individual radioisotope such as ^{14}C have variable energies over a limited range of values. The more energetically are the electrons emitted, the easier is the radioisotope to detect, but the more damaging is the radiation to a living organism. The longer the half-life of a radioisotope, the longer will it continue to emit a significant quantity of radiation if it is absorbed into tissues. These considerations of safety for the experimenter do not always have a large part in deciding which radioisotope to use as a tracer. If you want to study the incorporation of carbon from CO_2 into organic molecules then you have to use ^{14}C, and incorporation of inorganic phosphorus must be followed with ^{32}P. Cost is another factor;

^{3}H- or ^{32}P-labelling is generally cheaper than ^{14}C-labelling. The drawback to using tritium is that its emission of electrons is so weak that these electrons can only be measured quantitatively by scintillation counting at low efficiency (see below).

11.2 Scintillation counting

In a **scintillation counter** the radioactive material is mixed with a scintillation fluid. This is an organic solvent in which are dissolved other organic molecules, called **scintillants**. When one of these molecules is hit by an electron the scintillant emits photons of light which are detected by a photoelectric cell in the counter. An electron of high energy (as from ^{32}P) causes a relatively bright flash of light to be produced, while a low-energy electron (as from ^{3}H) produces a much less bright flash. Not all the electrons emitted by a radioisotope have the same energy, and so, in the case of ^{3}H, many of the flashes of light from the scintillants are too faint to be detected by the photocell, and only a proportion of the disintegrations that have occurred are registered as counts. The ratio of measured counts per minute (cpm) multiplied by 100, to disintegrations (of radioactive atoms) actually taking place per minute (dpm) is the percentage efficiency of counting. Efficiency of counting with tritium is only about 15% at best, while ^{32}P can be counted at 90% or higher efficiency, and ^{14}C and ^{35}S fall between these limits at about 60% efficiency of counting. It is usual practice to correct measured cpm to dpm once the percentage efficiency of counting is known:

$$dpm = cpm \times 100/\% \text{ efficiency}$$

Unfortunately, the efficiency of counting a given isotope is not constant, even with the same scintillation fluid in the same counter. This is because the maximum attainable efficiency is very often diminished by the presence in the scintillation fluid of quenching agents (such as water or oxygen) which are in the sample that is to be counted. However, it is usually possible to establish the efficiency at which an individual sample has been counted.

How is this done? A scintillation counter can discriminate between flashes of light that have different intensities, over a scale range of 1 (lowest detectable level) to 1000. The effect of quenching is to lower the brightness of the flashes of light, and as quenching becomes greater the number of counts recorded in the higher levels will progressively become a lower and lower proportion of the total measurable counts. The scintillation counter may be set up to record counts over two limited ranges of levels, called channel A and channel B. For instance channel B may be set to be the range from level 1 to 600, and channel A from 450 to 600. Thus, channel B accumulates all the detectable counts, while channel A only records the most energetic disintegrations.

If now we take a standard sample (which can be bought) that is free of quenching agents and has a known dpm, the channel B cpm will allow us to determine the efficiency at which it can be counted, and we also note the ratio of channel A to channel B. A small amount of a quenching agent (such as CCl_4) has been added to the next standard sample and it is now counted. The A / B ratio will change and the total cpm (channel B) will be less because efficiency has decreased. By progressively counting standards with more quenching agent present we can establish the relation between efficiency and the A / B ratio. This relation will only be valid for the particular radio-isotope and scintillation fluid that have been used, and only between the upper and lower values of the A / B ratio that are established. With these limitations, the efficiency of counting any new sample can be found from its A / B ratio, and the channel B count can be corrected to dpm. When the A / B ratios for a whole series of samples are close together, it is usually sufficient to determine the efficiency for just one sample and use the same value for all the samples.

Tritium cannot be detected by a Geiger counter, and these instruments are nowadays rarely employed for quantitative work. Radioactive emissions will fog photographic film, and X-ray film is much used to reveal the positions of radioisotopes on chromatograms or gels.

The half-lives of ^{3}H (12.3 years) and of ^{14}C (5730 years) are so long that the decay of total radioactivity is negligible during experiments that last for only a few weeks. However, with ^{32}P and ^{35}S it is usually necessary to correct n_t dpm (measured t days after the start of an experiment) to the value (n_0) that n_t dpm would have represented at the start of the experiment. To do this we use the equation

$$n_0 = n_t \, e^{kt}$$

(see Chapter 5) where k is the decay constant (expressed in the same terms as t, i.e. day^{-1} in this case) of the radioisotope.

Units of radioactivity

The older unit, the curie (Ci), is the activity of 1 g of pure radium together with the radon (also radioactive) that is formed by the decay of the radium, and is 2.2×10^{12} dpm. This arbitrary unit has now been replaced by the becquerel (Bq), which is defined as 1 disintegration per second, which is the same thing as 60 dpm.

Hence, $1 \text{ Ci} = 3.7 \times 10^{10} \text{ Bq}$

Specific activity

This is essentially dpm mole^{-1} of radioactive material. However, it is often given as μCi mmole^{-1}, which is numerically not the same, but which most frequently is a reasonably small number (1 μCi mmole^{-1} is 2.2×10^6 dpm mmole^{-1}, which is equivalent to 2.2×10^9 dpm mole^{-1}).

For a given radioisotope there is a maximum attainable specific activity. Take the example of pure $^{14}C^{16}O_2$. One mole of this will contain initially 6.02×10^{23} atoms of ^{14}C. The half-life is 5730 years, so that $k = 0.693 \div 5730$ year$^{-1} = 1.209 \times 10^{-4}$ year^{-1}. Consequently, after 1 year, the number of ^{14}C atoms remaining will be $6.02 \times 10^{23} \div e^{0.0001209} = 6.019 \times 10^{23}$ atoms. This means that in 1 year, 1×10^{20} atoms disintegrate, and the number disintegrating in 1 minute will be approximately $1 \times 10^{20} \div (365 \times 24 \times 60) = 1.90 \times 10^{14}$. Thus, the highest specific activity that is theoretically possible in a molecule containing 1 atom of ^{14}C is 86 Ci mole^{-1}. As pure ^{14}C is not commonly produced, the specific activity will in practice be lower than this. Of course, the more atoms of ^{14}C that can be incorporated into a molecule, the greater will be the specific activity of that molecule. For instance the amino acid alanine could contain three atoms of ^{14}C per molecule.

Higher specific activities are possible with radioisotopes that have half-lives much shorter than ^{14}C. For instance, pure $Na_2{}^{32}PO_4$ could have a maximum specific activity of 9.2×10^6 Ci mole^{-1}.

The disadvantage of ^3H-labelling (which is that the efficiency of counting is low) is offset by the higher specific activities that are attainable (relative to ^{14}C) and by the fact that many organic molecules contain several atoms of hydrogen, more than one of which may be replaceable by tritium.

Double labelling

A single molecule can contain two different radioactive elements. For example benzylpenicillin could be prepared with ^{3}H- and ^{35}S-labelling. Such labelling may allow one to establish at what stage of metabolism a molecule is broken into separate parts. The two radioelements must emit electrons with sufficiently different energies (e.g. ^{3}H and ^{14}C, or ^{14}C and ^{32}P) that allow each element to be counted in separate channels of a scintillation counter. In principle, it is also possible to count two elements that emit electrons with similar energies (e.g. ^{14}C and ^{35}S) if their two half-lives are very different. This method is rarely used in practice.

Pulse labelling

One may wish to add a radioisotope to an experimental system for only a short time – minutes or seconds rather than hours – for example to find what are the first organic molecules that become labelled when $^{14}CO_2$ is fixed during photosynthesis. The brief exposure may be achieved by stopping the reaction very soon after adding the radioactive substrate (by heating or by adding an inhibitor), or by adding a large excess of the unlabelled substrate, so that the incorporation of radioactivity is drastically decreased.

Background counts

Any counting apparatus will record some small number of counts even when no radioactive material is present in the sample being counted. Such spurious counts are called **background**, and in a scintillation counter may have several possible causes: cosmic radiation, hypersensitivity of the detector, luminescent material in the sample, spilled radioactive material inside the counter. These last two causes are problems that can be avoided and must be resolved before reliable counting can be done. The 'unavoidable' background in a scintillation counter is about 20 cpm, and the background count has to be subtracted from every measured count to get a correct estimate of the count that is really due to radioactivity in the sample. It is very important to have a statistically reliable measure of the background (see below).

Self-absorption

When a radioactive substance is in solid form, rather than in solution, some of the emitted electrons may have insufficient energy to escape out of the solid, so that they will not be counted. This happens particularly with tritium. Bacteria that are labelled internally (e.g. with ^{3}H-thymine in their DNA) will show a lower count while they are intact than when the organisms are disrupted and the DNA passes into solution. The same phenomenon can occur with other isotopes that emit more energetic electrons if the radioactive solid particles are bigger.

11.3 Statistics of counting radioisotopes

If you had just ten atoms of ^{14}C you might well wait for more than 500 years before observing the decay of one atom, though you might possibly observe one to decay in the first minute. All you can be fairly sure of is that about five of the ten atoms of ^{14}C will have decayed after 5730 years. On the other hand, if you started with 6×10^{13} atoms of ^{14}C you would be reasonably certain of seeing close to 2×10^4 dpm. What this means is that low counts have low reliability, and may show a large percentage error from the true count, whereas large counts have small percentage errors. The probable error of a single estimate is equal to the square root of the number of counts. A count of 100 dpm may have an error of 10%, while a count of 10 000 dpm may have an error of only 1%.

If you measured a background count for just 1 min, you might get any value between 15 and 25 counts, when the true average value was 20 cpm, that is, an error of at least 25%. A better procedure would be to count background for 100 min, when about 2000 counts would be recorded, and the probable error would be only about 2%. Better still is to count background for 10 min and repeat the count ten times. The mean count and its standard deviation then can be found. In general, if a sample has a low activity that needs to be determined accurately (often it does not; see below), the sample should be measured several times, until a reasonably large total number of counts have been accumulated, from which the mean can then be calculated.

Very high rates of counting

In any radioactive decay, there is always a possibility that two disintegrations may occur so closely together (in time) that the counting apparatus records the double event as only a single count. Such overlaps will obviously take place more and more often when increasingly large numbers of disintegrations are going on during a short interval. Consequently, very high cpm values are likely to be underestimates of the true activity. For greater accuracy, a very high cpm should be repeated with a smaller amount of the same radioactive material.

The reality

In spite of what has just been said above, there is, more often than not, no need to repeat low or high counts. In many experiments, all that one wants to know is which, out of several samples, contains a high level of radioactivity. Suppose you want to find out which of five samples are radioactive, and after counting each for 1 min you get values of: 55, 6505, 3464, 200, 51 counts. It is obvious that samples 2 and 3 contain nearly all the radioactivity, and whether the values for samples 1, 4 and 5 are accurate (even ±20%) will not matter. Likewise, it will frequently be unnecessary to determine the counts of samples 2 and 3 more precisely.

12 | Growth in batch cultures

. . . did show that at that time there was 4000 persons derived from the very body of the Chiefe Justice. It seems the number of daughters in the family having been very great, and they too have most of them many children and grandchildren and great-grandchildren. This he tells me as a most known and certain truth.

Samuel Pepys

A **batch culture** is grown in a closed system. The medium may be solid or liquid, contained in a Petri dish, test tube, flask or fermenter, and may or may not be accessible to sterile air. Samples may be taken from the culture at intervals. However, there is no *continuous* addition of fresh medium with a corresponding *continuous* removal of an equal volume of spent medium containing organisms. This latter procedure is continuous culture, and will be considered in the next chapter.

Usually the inoculum will be a pure culture, that is, organisms believed to be all of the same kind, such as *Escherichia coli* or *Staphylococcus epidermidis*. The object will be to study some property of a particular strain. Sometimes the inoculum may be an unknown mixture of organisms, like soil or pus. In these cases, the object will be to find out what organisms were present in the inoculum, and perhaps to go on to isolate some of these as pure cultures.

12.1 Assessment of growth

Often all that is wanted is a 'yes' or a 'no' answer. Looking at a culture after incubation is then enough. Colonies appear (or do not) on a solid medium, or a liquid medium becomes turbid (or does not). Getting quantitative results calls for more effort.

Viable counts

One inoculates a solid medium with a small volume taken from a culture and records how many colonies are formed after incubation. There are difficulties.

It is important to get a suitable number of colonies on a Petri dish – too few (say fewer than 20) and the sampling error will be significant; too many (say more than 200) and counting may be impossible because neighbouring colonies fuse together. For this reason it is necessary to use several different dilutions of the culture on separate dishes. Colonies appear only after some hours of incubation, and the colonies then have to be counted. The whole process is expensive in terms of materials and time. The medium and the conditions of incubation must be suitable for the organisms that are being counted. Non-appearance of colonies does not prove that the organisms in the inoculum were dead – they might have grown under altered circumstances. A single colony does not always represent a single organism in the inoculum: a yeast with several buds, or a cluster of staphylococci or a chain of streptococci will yield only one colony.

The great advantage of viable counts is that they can measure very sparse populations that would be unrecognisable by other methods.

Total counts

These are done with a counting chamber. This is a glass slide on which is engraved a series of parallel lines, each 1 / 20 mm from the next. A second set of parallel lines, also 1 / 20 mm apart, is also engraved at right angles to the first series, to create a grid of squares, each of which has an area of $1 / 400$ mm^2. A drop of the suspension of organisms is put on the grid and covered with a special glass slip, which traps a film of the suspension, 1 / 5 mm deep. The number of organisms seen (by phase-contrast microscopy) inside one square is the number in a volume of $1 / 20 \times 1 / 20 \times 1 / 5$ mm^3. This is the same as $1 / 2\,000\,000$ ml. Consequently, the **count per square must be multiplied by 2×10^6** to yield the number of organisms ml^{-1}. To get an accurate answer the average count per square should be based on the counts of about 100 squares derived from several fillings of the chamber because the thickness of the trapped film above the grid is variable. The method is quicker than the viable count, but has the disadvantage that any suspension containing less than 2×10^6 organisms ml^{-1} will give a count of less than 1 organism per square – in other words sparse populations cannot be counted. Clusters and chains of organisms are again a problem.

The **viable** and **total counts** are both called **direct methods** because their result is the number of organisms (or clusters) per unit volume. **Indirect methods** make use of some property of the culture that is proportional to the number of organisms per unit volume.

Of these, **measurement of turbidity (optical density)** is the most widely used. It has the advantages of being quick and easy, and is accurate within limits. There is a lower limit below which no turbidity can be detected (about 1×10^7 organisms ml^{-1}) and an upper limit where the turbidity becomes too great (about 5×10^9 organisms ml^{-1}) and further increase cannot be measured, though a dense suspension can, of course, be diluted and then measured.

Microorganisms of different species may have different sizes or shapes; even the organisms from a pure culture may change in these respects when grown under altered conditions. Hence, the relation between optical density and the concentration of a suspension of organisms can also be quite variable, and needs to be found by experiment.

A dense suspension of the organisms is grown under appropriate conditions, diluted and counted microscopically. Various other dilutions of the suspension are made, with known numbers of organisms ml^{-1}, and the optical densities of these suspensions are measured. The resulting data allow a calibration curve to be drawn, which relates number of organisms ml^{-1} (x axis) to optical density (y axis).

12.2 Phases of growth

Lag phase

For a period of time after inoculation there is no substantial increase in the number of organisms ml^{-1}. This lag is attributed to the need by the organisms in the inoculum to adjust their metabolism (form new enzymes) for biosynthesis in the new medium. How long is the lag phase? Even when the organisms of the inoculum have been actively growing in a medium of the same composition there is still a short lag, perhaps 1 or 2 hours. The lag will be longer if the inoculum came from a stored refrigerated culture.

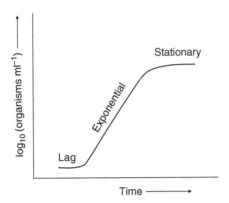

Fig. 12.1 Phases of growth in a batch culture.

Organisms taken from a rich medium may lag for 6 hours or more if they are put into a minimal medium. If the inoculum is small, e.g. 1×10^3 organisms (ml of medium)$^{-1}$, then the duration of the lag phase will be greatly overestimated by any method that fails to detect sparse populations. For example if optical density were used there would be no discernible growth until a 10 000-fold increase in the number of organisms ml^{-1} had really taken place. In general, growth will already be in the exponential phase when turbidity is first seen.

Exponential phase

This is most easily recognised (and most easily defined) as the period during which a graph of **the logarithm of the number or weight of organisms per unit volume** (or the logarithm of a function directly related to this, such as optical density), **plotted against time, yields a straight line of positive slope**. This comes to the same thing as saying that it is the period during which the rate of increase of the population at a given moment is directly proportional to the size of the population at that moment. As Fig. 12.1 shows, most of the growth of a batch culture happens during the exponential phase. How long is the exponential phase? This will depend on the initial number of organisms ml^{-1}, their rate of growth and on the maximum population that the medium can produce.

To understand calculations that relate to the exponential phase you need a knowledge of logarithms. See Chapter 5 for a fuller description. Two important numbers apply to the exponential phase: the **doubling time** (or mean generation time) (**t_d**) which is constant during the exponential phase, and may have units of minutes or hours; and **the specific growth rate (μ)** which also is constant, and has units of a pure number per minute or per hour. (When multiplied by 100, μ is equal to the percentage increase per unit time.) The relation between μ and t_d is:

$$\mu = 0.6931/t_d$$

The derivations of t_d and μ are described in Chapter 5.

Stationary phase

Eventually some condition (depletion of a vital nutrient, change of pH value in the medium, etc.) stops the exponential increase and the rate of growth decreases until it becomes zero. The culture then is in the stationary phase. The duration of this phase depends very much on the properties of the organisms that have grown: some species can remain for several days with little loss of numbers or viability, others enter a decline phase very soon after growth stops, and their viability quickly falls. If the organisms lyse there will be a decrease of turbidity.

Although growth has stopped, other metabolic activity may occur in the stationary phase. In particular, it is at this stage that many fungi produce secondary metabolites, notably antibiotics, and some bacteria (especially auxotrophs when they have depleted the medium of an essential growth factor) become excretors of amino acids. The bacteria (of the genera *Bacillus*, *Clostridium* and *Sporosarcina*) that can make endospores do so to a limited extent at all stages of batch culture, but it is during the stationary phase that there can be massive sporulation in response to starvation, when almost every organism may make a spore.

12.3 Microbiological assays

Growth in a batch culture will stop if an essential nutrient is used up before other conditions become limiting. For example, *Lactobacillus casei* needs folic acid, and will fail to grow if this vitamin is absent from a medium that is adequate in other respects. There will be a range of suboptimal concentrations of folate over which the extent of growth (numbers of bacteria or

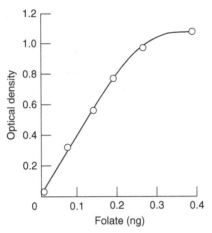

Fig. 12.2 Growth of *Lactobacillus casei* in a medium (5 ml) containing increasing amounts of folic acid. Cultures were incubated anaerobically at 37 ° C for 24 h before turbidity was measured.

turbidity) will be proportional to the concentration of folate in the medium. There will also be a concentration of folate enough to support the heaviest growth that the medium can sustain, and higher concentrations will then cause no further increase in growth (Fig. 12.2).

A solution containing an unknown concentration of folic acid can be assayed by taking a series of test tubes containing a medium for *L. casei* that is without folate and adding various volumes of the test solution to different tubes. After inoculation and incubation the growth in each tube is measured. If there is no folic acid (at a detectable concentration) in the test solution there will be no growth in any tube of the assay. When the test solution does contain folate then some tubes, which received excess folate, will show maximum growth. The object is to find one or more tubes in which growth is at an intermediate level, so that the concentration of folate in the assay tube can be read from the standard curve. The concentration in the test solution can then be calculated.

While the principle of microbiological assays is simple, the practice is not. The method is so sensitive that all apparatus must be very clean, and components of the medium must be free of the substance that is being assayed. To avoid growth in the blank it is often necessary to grow the inoculum in a suboptimal concentration of the growth factor, so that the organisms carry none into the assay medium. Reproducibility is never very good, repeated assays of the same material vary by 10% to 20%, and a series of standards has to be part of every assay.

Inhibitors of growth

Sometimes growth is inhibited when you didn't want it to be. Reasons could be:

mistake(s) in preparing the medium (especially wrong pH value);
impurity in a reagent;
non-viable inoculum;
inoculum may be a contaminant that does not grow in your medium;
inhibitor (e.g. traces of detergent) on the inner surface of the culture vessel (cotton wool or rubber plugs can exude inhibitors, especially when the plugged vessel is sterilised with dry heat).

Carefully making fresh medium in really clean glassware, and checking the inoculum microscopically will usually solve these problems. However, much time can be lost in tracking down the cause of a failure. It's fortunate that Fleming took the trouble to find out why a mould was inhibiting growth of staphylococci in one of his Petri dishes.

More commonly, one is hoping to see inhibition, in order to recognise a compound that is active against the growth of an organism, or else to establish the lowest concentration of a known inhibitor that is effective against a particular organism.

In a hospital laboratory it is important to find which antibiotic will inhibit an infective organism that has been isolated from a patient (Fig. 12.3). A greater number of potential inhibitors can be tested with a micro-titre plate, where the different compounds are added to wells (up to 96) which all contain the test organisms in a growth medium. Similar procedures are used in industry when searching for new antibiotics that are to be tested against various pathogens.

The concentration of a solution of an inhibitor can be determined by zone assay (Fig. 12.4). This method is valuable when purifying a new compound of unknown structure, when no other means of assay may be available.

Determining the **minimum inhibitory concentration (MIC)** of an anti-bacterial compound against a specific organism is technically simple. One

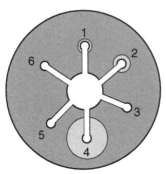

Fig. 12.3 Determination of the susceptibility of a bacterial culture to various inhibitors. A disk of filter paper has prongs, the ends of which are impregnated each with a different microbial agent, 1 to 6. The disk is put onto an agar medium that has been seeded with the organism: after incubation, clear zones will appear where a successful inhibitor has diffused into the agar. In this example, the organisms are most sensitive to compound 4, while 1 and 2 are weaker inhibitors, and 3, 5 and 6 are ineffective.

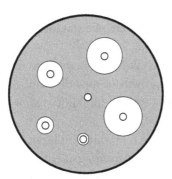

Fig. 12.4 A zone assay of an inhibitor. Molten medium containing agar is seeded with an organism that is sensitive to the inhibitor. The medium is poured into a dish, and when it has set holes are cut out. Usually, each of these wells can hold 200 μl of a solution, and various amounts of the inhibitor (each in 200 μl) are next pipetted into different wells. After incubation, clear zones are seen, against the background of growth, round wells that contained sufficient amounts of the inhibitor. The relation can then be found between the diameters of the clear zones and the amounts of inhibitor in the wells. A solution (200 μl) of the same inhibitor (but of unknown concentration) can be assayed by finding the diameter of the clear zone that it produces and relating this to the standards.

can set up a series of test tubes containing medium with increasing concentrations of the compound. After inoculating with the test organisms and then incubating the tubes one observes what is the lowest concentration of the inhibitor that has prevented growth. Nevertheless, the MIC of a given inhibitor to a given organism does not necessarily have a constant value. Any of these factors may alter the MIC:

composition of the medium (minimal or complex);
conditions of incubation (aerobic, anaerobic, temperature);
length of time of incubation;
size of inoculum.

The **minimum effective dose** to protect an animal or plant from infection has to be determined by further experiments, and may be considerably different from the MIC.

12.4   Virus multiplication

The number of virus particles in a suspension *can* be found by electron microscopy, though this is very laborious. More usually the suspension is spread onto a lawn of a susceptible host (indicator) organism, which may be a tissue culture (for animal or plant viruses) or agar medium seeded with a bacterial culture (for bacteriophages). A virus infects its host and causes necrosis or lysis, so releasing more viruses. These in turn cause further cycles of infection and multiplication. The result is that after a suitable time of incubation a hole (called a **plaque**) will appear against the background of growth of the indicator. Each plaque, though it will contain many millions of virus particles, will represent one virus that was present in the suspension put onto the lawn.

From its definition as an intracellular parasite, it follows that a virus will only multiply when a host is available. The **single burst experiment** illustrates the process (Fig. 12.5). Phage and host bacteria are mixed in $1:1$ proportions. After 5 min unattached phages are inactivated by adding an antiserum. The infected bacteria are at once plated at various dilutions onto a lawn of the same bacteria as indicator. Each plaque that later appears from these first platings will represent one infected bacterium (no matter how many phages attacked it), and the height (a) in the figure is the total number of infected bacteria. Sampling and plating of the infected culture are

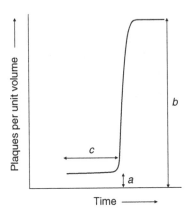

Fig. 12.5 The single burst experiment.

repeated frequently. For a time, the latent period (c), there is no change in the number of plaques per unit volume. Then lysis of the bacteria occurs and many phages are released in a brief time, and the number of plaques increases markedly and quickly. When all the bacteria have lysed the number of plaques becomes constant (b) because no hosts remain to produce more phages.

The size of the burst of new phages is $b - a$, and the number of phages released by one infected bacterium is $(b - a) / a$. For example, if $a = 10$ and $b = 210$, then the individual burst size would be $(210 - 10) / 10 = 20$. The burst size per organism is rather constant for a given organism and phage, but differs considerably between other pairs of host and virus, in the range between 10 and 50.

Lysogeny

The DNA of a phage may be incorporated into the chromosome of a host bacterium without causing lysis. Every daughter cell of this lysogenised organism will carry the DNA of the phage in its genome. In this way 2^n copies of the phage DNA will be formed after n generations of exponential growth of the host. Certain agents (e.g. UV light) may induce a lytic response in a lysogenised organism. The phage genome is excised from the bacterial chromosome and infective phages are produced inside the host, which then lyses to liberate several infective phages.

If one bacterium were lysogenised and then grew for 10 generations by binary fission there would be 2^{10} (1024) lysogenised bacteria. Should this population then be induced to give the lytic response, and each bacterium produced 10 phages, then 10 240 new viruses would have originated from the initial infection by one phage.

13 | Growth in continuous culture

Dis-moi ce que tu manges, je te dirai ce que tu es.

Brillat-Savarin

Of all the topics covered in this book, continuous culture puts the greatest strain on mathematical ability. There are here several equations that are not easy to derive, but must be known because they may not be presented to you in an examination question. These equations really have to be learned by heart. Fortunately, it is not very difficult to put numerical values into the equations to produce the answers.

There are two modes of continuous culture, the **turbidostat** and the **chemostat**.

13.1 The turbidostat

The turbidostat is illustrated in Fig. 13.1. The culture vessel contains medium that can be stirred, aerated and kept at constant temperature and pH value (by adding acid or alkali) as needed. The medium is inoculated and allowed to grow enough to become turbid. Pump 2 then circulates a small part of the culture through a photometer which measures this turbidity. The output from the photometer goes to a control unit which activates pump 1 whenever the turbidity rises above some set value, so that fresh sterile medium enters the culture vessel. The volume of the growing culture remains constant owing to the overflow tube, and the turbidity settles to the set value.

The turbidostat allows organisms to be grown at their maximum rate in any particular medium. However, it is difficult to operate successfully, chiefly because organisms tend to stick to the surfaces of the culture vessel and sampling loop, and so cause false readings of the photometer.

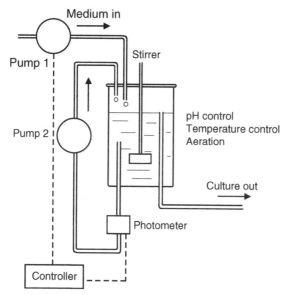

Fig. 13.1 The turbidostat.

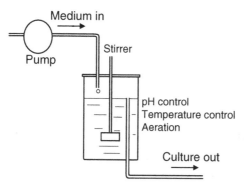

Fig. 13.2 The chemostat.

13.2   The chemostat

The chemostat is more widely used, and all of what follows relates to this form of continuous culture. This is a less elaborate system than the turbidostat (see Fig. 13.2), although the real apparatus looks very much more complex than the simplified figure may suggest. When the culture has grown to a discernable turbidity sterile medium is added at a constant rate, typically as one-fifth of the culture volume per hour. The turbidity of the culture settles to a constant level, provided that the rate of addition (the dilution rate) is not so high as to wash out the organisms from the culture vessel in the flow of sterile medium. The rate of growth of the culture precisely matches the dilution rate, and this steady state can often be maintained for several weeks. This balance between dilution and growth seems obvious, and can be demonstrated mathematically.

After a given interval of time (t hours) the number of organisms in the culture at time zero (n_0) will tend to be lowered by dilution with fresh medium by a factor of n_0 / e^{Dt} where D is the dilution rate (fresh medium added per hour expressed as a fraction of the culture volume). At a steady state this dilution must be exactly matched by the growth of new organisms during time t, which will be $n_0 e^{\mu t}$ where μ is the specific growth rate (see Chapter 5: Logarithms) per hour. Hence we can write:

$$n_0 e^{\mu t} - n_0 / e^{Dt} = 0$$

at a steady state.

Dividing by n_0 we have

$$e^{\mu t} - 1/e^{Dt} = 0$$
$$\text{and so } e^{\mu t} = 1/e^{Dt}$$

Multiplying both sides by e^{Dt} we get

$$e^{\mu t} \times e^{Dt} = 1 = e^{(\mu t + Dt)}$$

Therefore $\mu t + Dt$ must $= 0$ (see Chapter 5: Logarithms) and $\mu t = -Dt$ and (dividing both sides by t) $\mu = -D$. The rate of dilution and the rate of growth are numerically equal, but operating in opposite directions.

When organisms are grown in a complex medium it is not possible to be sure what factors are limiting growth and leading to the establishment of a steady state. However, in a simple chemically defined medium **one can supply an essential nutrient** (e.g. carbon, nitrogen, phosphorus or sulphur) **at a low concentration which limits the population density**.

The rate of growth of the culture is determined by the dilution rate

Suppose we have a culture **at a steady state** which contains n_c organisms in the culture vessel. If the dilution rate is D h^{-1} then after t hours the total

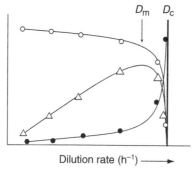

Fig. 13.3 Events during growth of bacteria in a chemostat. ○, dry weight of organisms ml^{-1}; ●, concentration of growth-limiting substrate in culture vessel; △, yield of organisms in outflow (mg dry weight h^{-1}). D_m is the dilution rate that gives the maximum yield; D_c is the critical dilution rate above which wash-out occurs.

number of organisms in the vessel plus those in the outflow (n_t) will be n_c + $n_c Dt$ (the concentration of organisms in the outflow is the same as in the vessel at steady state), i.e. $n_t = n_c Dt + n_c$. At a fixed dilution rate both n_c and D are constant, so that $n_c D$ is also a constant (C), and therefore $n_t = Ct + n_c$. Hence, a graph of n_t against t will be a straight line of slope (dn_t/dt) equal to $n_c D$ and intercept n_c on the n_t axis. This shows that **at steady state the organisms in the vessel are *not* growing exponentially.** (Exponential growth does *not* lead to a straight-line graph unless log n_t is plotted against time.) They should not to be likened to organisms in a batch culture that are in the exponential phase; rather, they resemble organisms in the retardation (post-exponential) phase of a batch culture. However, unlike a batch culture, all the organisms in the chemostat remain in this same stage for the whole period during which steady-state conditions are maintained.

Relations between the various properties of a continuous culture of bacteria in a chemostat are shown in Fig. 13.3. At zero dilution rate the organisms grow to the highest concentration that is permitted by the limiting substrate, which itself is depleted to a very low concentration. As the dilution rate is increased the concentrations of organisms and of substrate change little, though the yield of organisms per hour rises (because progressively greater volumes of outflow are being generated with little change in the concentration of organisms therein). Only when the dilution rate approaches the wash-out limit do the concentrations of organisms and of substrate begin to alter quickly. The chemostat allows a choice of bacterial density (by fixing the concentration at which the limiting nutrient is supplied) and rate of growth (by regulating the dilution rate), and makes available copious amounts of organisms that have all been produced under the same conditions. **These variable factors, and particularly altered nutrient limitation** (e.g. going from phosphate to magnesium limitation) **can lead to pronounced changes in the chemical composition of the organisms.**

The conditions in a chemostat are very strongly selective for organisms that can grow more heavily or faster than the original population. A contaminant, or a faster-growing variant, may take over the culture. Recombinant plasmids tend to be lost because organisms without plasmids grow more quickly.

13.3   Growth rate and concentration of the limiting substrate

The dilution rate and the concentration of the limiting nutrient in the culture fluid within the vessel (not in the medium that is pumped in) are related:

$$\mu = \mu_{max}[s/(K_s + s)]$$

where μ is the growth rate (= absolute value of D), μ_{max} is the maximum growth rate (when the substrate is not limiting), s is the concentration (moles or g L^{-1}) of limiting substrate *in the culture vessel*, and K_s is the saturation constant, numerically equal to the growth-limiting substrate concentration (moles or g L^{-1}) that allows growth at half of its maximum rate. (Compare the Michaelis–Menten equation in Chapter 8: Enzymes.) The equation can be arranged as:

$$K_s = s(\mu_{max} - D)/D$$

The values of K_s and μ_{max} can be calculated by measuring the two different steady-state values of s that result from two different (known) dilution rates:

$$K_s = s_1(\mu_{max} - D_1)/D_1 = s_2(\mu_{max} - D_2)/D_2$$

Solving $s_1(\mu_{max} - D_1) / D_1 = s_2(\mu_{max} - D_2) / D_2$ for μ_{max} then allows K_s to be found.

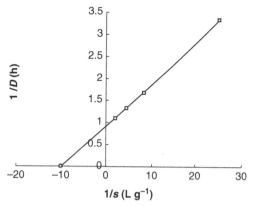

Fig. 13.4 Graphical determination of K_s and μ_{max} from plots of $1/s$ and $1/D$ ($= 1/\mu$). The intercept on the $1/D$ axis (here approx. 0.9) is equal to $1/\mu_{max}$ so that μ_{max} is $1.1\ h^{-1}$. The intercept on the $1/s$ axis (here approx. -10) is equal to $-1/K_s$ so that K_s is $0.1\ g\ L^{-1}$.

Another way of determining μ_{max} and K_s is to plot $1/D$ against $1/s$ (see Fig. 13.4) for several values of D and the corresponding values of s. (Again compare with the Lineweaver–Burk plot, Chapter 8: Enzymes.)

Rate of uptake of nutrients

In steady-state conditions

$$\mu = Yq \text{ or } q = \mu/Y$$

where q is the **rate of uptake of the growth-limiting substrate** (moles (or g) per gram of organisms per hour), and Y is the **yield value** (grams of organisms formed per mole (or g) of limiting substrate consumed). The value of Y can be found from the relation:

$$x = Y(S_r - s) \text{ or } Y = x/(S_r - s)$$

where x is the concentration of organisms in the culture vessel, S_r is the concentration of the limiting substrate in the entering medium and s is its concentration in the culture vessel. Knowing Y and μ ($= D$) we can determine q. In practice Y proves not usually to be constant (even for a given organism in a given medium) but varies with the growth rate.

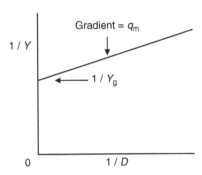

Fig. 13.5 Graphical determination of Y_g and q_m.

In carbon-limited cultures we may assume that a part of the substrate must be oxidised to provide energy for maintenance functions, such as solute gradients, motility and repair of macromolecules. If this maintenance energy does not vary with growth rate then the **rate of substrate uptake for maintenance (q_m)** can be estimated by extrapolation to zero growth rate of a plot of q against D (not illustrated), which should give a straight line with an intercept equal to q_m on the ordinate (vertical axis).

When q_m is known one can determine q_g (the substrate requirement associated solely with growth) from the relation:

$$q_g = q - q_m$$

Re-arranging and dividing through by μ gives $q/\mu = q_g/\mu + q_m/\mu$. Now, $q/\mu = 1/Y$ (see above), so that

$$1/Y = 1/Y_g + q_m \times 1/\mu$$

where Y_g is the yield value attributed to growth alone. **A graph of $1/Y$ against $1/D$ ($=1/\mu$) should produce a straight line of gradient q_m and intercept on the ordinate of $1/Y_g$** (see Fig. 13.5).

Mathematical treatment of the chemostat can be taken to a much higher level than it is here.

13.4   A sample problem

A culture of an unidentified bacterium was grown in a chemostat that had a culture volume of 200 ml with glucose as the growth-limiting nutrient. The

Table 13.1 *Results of hourly assays of bacterial concentration (x) and glucose concentration (s)*

Time (h)	x	s	Time (h)	x	s
0	0.48	0.04	10	0.27	0.46
1	0.48	0.04	11	0.262	0.476
2	0.48	0.04	12	0.26	0.48
3	0.46	0.08	13	0.26	0.48
4	0.45	0.10	14	0.33	0.34
5	0.445	0.114	15	0.36	0.28
6	0.44	0.12	16	0.38	0.24
7	0.44	0.12	17	0.387	0.226
8	0.35	0.30	18	0.39	0.22
9	0.30	0.40	19	0.39	0.22

concentration of glucose in the medium supplied to the culture vessel was 1.0 g L^{-1} and the saturation constant (K_s) for glucose with these organisms was known to be 0.1 g L^{-1}.

The culture was supplied with medium at a constant flow rate of 60 ml h^{-1} for several hours and after sampling for a further 2 h the rate was changed to 120 ml h^{-1}. Further changes in flow rate were made after 7 h (from 120 to 182 ml h^{-1}) and after 13 h (from 182 to 150 ml h^{-1}). Samples of the effluent were assayed hourly for bacterial concentration (x g dry wt bacteria L^{-1}) and for glucose (s g glucose L^{-1}). Results are shown in Table 13.1.

(1) Plot the data and describe with *brief* explanations the changes which accompany the alterations in flow rate.
(2) Define, with units, the term dilution rate, and calculate the dilution rates operating at different times in this problem.
(3) Calculate from the steady-state data the average values for:
  (a) the yield constant (Y) and give its units;
  (b) the growth rate constant (μ_{max}) and give its units.
(4) Calculate the output of bacteria (g h^{-1}) for the final steady state.
(5) Predict the steady-state concentrations of bacteria and of glucose that would be found at a flow rate of 90 ml h^{-1}.

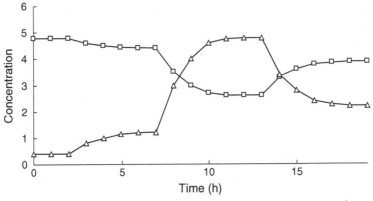

Fig. 13.6 $\triangle$, s = glucose (g L^{-1}); $\square$, x = organisms (g dry wt L^{-1}).

Answers

(1) See Fig. 13.6. The culture is in a steady state initially. When the dilution rate is changed (up or down) the culture takes about 3 h to settle to a new steady state. At the lowest dilution rate the concentration of substrate is also at its lowest, while the concentration of organisms in the culture is highest. At the highest dilution rate the concentration of substrate is also at its highest, while the concentration of organisms in the culture is lowest.

(2) **Dilution rate** is the ratio of the volume of medium entering the culture vessel per unit time (usually per hour) to the fixed volume of culture in the vessel.

Time (h)	Dilution rate (h^{-1})
0–2	0.3
2–7	0.6
7–13	0.91
3–19	0.75

(3)(a) $Y = x/(S_r - s)$ and $S_r = 1.0\,\mathrm{g\,L^{-1}}$

When $x = 0.48$ g L^{-1} and $s = 0.04$ g L^{-1} then $Y = 0.48 / (1.0 - 0.04) = 0.48 / 0.96 = 0.5$. **The units are g bacteria per gram glucose.** (The mol. wt of glucose is 180 and so Y can also be stated as 90 g bacteria per mole of glucose.)

The other three steady states all give the same value (0.5 g bacteria per g glucose).

(3)(b) $\mu = \mu_{\text{max}} [s / (K_s + s)]$ and $K_s = 0.1 \text{ g L}^{-1}$

Replacing μ by D we can write $D / \mu_{\text{max}} = s / (K_s + s)$. Hence, $s \times \mu_{\text{max}} = D(K_s + s)$ and $\mu_{\text{max}} = D(K_s + s) / s$

When $D = 0.3$ $s = 0.04$, so that $\mu_{\text{max}} = 0.3 \times (0.1 + 0.04) / 0.04 = 0.042 / 0.04 = 1.05 \text{ h}^{-1}$

When $D = 0.6$ $s = 0.12$, so that $\mu_{\text{max}} = 0.6 \times (0.1 + 0.12) / 0.12 = 0.132 / 0.12 = 1.10 \text{ h}^{-1}$

Similarly when $D = 0.91$ $\mu_{\text{max}} = 1.10 \text{ h}^{-1}$, and when $D = 0.75$ $\mu_{\text{max}} = 1.09 \text{ h}^{-1}$

Average $\mu_{\text{max}} = 1.09 / \text{h}$

(4) At the final steady state $D = 150 \text{ ml h}^{-1}$ $(= 0.15 \text{ L h}^{-1})$ and there are 0.39 g bacteria L^{-1}. Hence the **rate of production of bacteria** $= 0.15 \times 0.39 = 0.0585 \text{ g h}^{-1}$.

(5) At a steady state of $D = 90 / 200 = 0.45 \text{ h}^{-1}$, $K_s = 0.1 \text{ g L}^{-1}$ and $\mu_{\text{max}} = 1.09 \text{ h}^{-1}$. We need to find the values of s (g glucose L^{-1}) and of x (g bacteria L^{-1}). By rearranging the equation $\mu = \mu_{\text{max}} [s / (K_s + s)]$ we can write

$$s = D \times K_s / (\mu_{\text{max}} - D)$$

Thus, $s = 0.45 \times 0.1 / (1.09 - 0.45) = 0.045/0.64 = \textbf{0.070 g glucose L}^{-1}$

$$x = Y(S_r - s)$$

$Y = 0.5 \text{ g g}^{-1}$, $S_r = 1.0 \text{ g L}^{-1}$ and $s = 0.07 \text{ g L}^{-1}$, so that $x = 0.5 \times 0.93 = 0.47 \text{ g L}^{-1}$.

14 | Microbial genetics

Some day perhaps you will enlighten me about the earthshaking significance of the double helix, etc. If it hadn't been worked out on a Tuesday, it would have happened in some other laboratory on Wednesday or Thursday.

C. M. MacLeod

Here computers do most of the data-handling. Highly sophisticated programs from the Internet can examine the nucleotide sequences of many genes, and deduce the amino acid sequences of proteins. Learning to use these programs needs hands-on practice in front of the screen. Microbiologists of the old school may deplore (= envy ?) the limited amount of quantitative work that the researcher now has to do. In this chapter we shall consider some topics that still are feasible with pencil and paper and a bit of thought.

14.1 Composition of DNA

The **discovery** of DNA was reported by Friedrich Miescher in 1871. Understanding the detailed **structure** of the DNA molecule became very important after Avery and Macleod had shown that DNA was the material of the gene. Each of the two strands in a molecule of DNA is a linear polymer of units of 2-deoxyribose 5-phosphate. Every one of these deoxypentose units is linked to a purine base (adenine (A) or guanine (G)) or to a pyrimidine base (thymine (T) or cytosine (C)). The sequence in which these bases occur (on the chain of deoxypentose phosphate units) is the genetic code. By careful analyses Erwin Chargaff found that equimolar quantities of A and T are present in double-stranded DNA; G and C are also equimolar, but A (and T) will not usually be equimolar with G and C (Chargaff's rules). Like Colin Macleod, Chargaff had good reason for a lack of enthusiasm about the attention given to Watson and Crick – 'our mass-media substitutes for saints' – Chargaff had the sharpest tongue in biochemistry!

There are various methods of determining the **%GC** in a given DNA. If for example the %GC were 32 this tells that of all the four bases (A, C, G and T) in the DNA, then 32% (in molar terms) are G and C. Hence 68% are A and T, and from Chargaff's rules it follows that 16% are C, 16% are G, 34% are A and 34% are T. Comparison of the %GC in DNA from different bacterial cultures can give an indication about their possible relatedness: if the %GC values differ by more than about 10 then there *cannot* be long identical sequences of deoxynucleotides (bases) in the two DNA samples, and so the two strains cannot be closely related. Similarity of %GC values from two cultures is consistent with their relatedness, but in no way does it prove relatedness because many differently ordered sequences of bases could lead to the same %GC value.

14.2   Parent strains and mutants

A wild-type strain is a pure culture isolated from its natural environment, whatever that may be – soil, water, your mouth, a septic wound. Such a culture can be the parent strain from which mutants are derived in the laboratory. (If you accept the 'theory of evolution by natural selection' then every contemporary parent strain must itself be a multiple mutant of ancestors that had different features.) Mutants are produced by some method that causes a change (mutation) in the genome of the parent which is inherited by subsequent generations of the mutant. Some of the mutations lead to a recognisable alteration of a parental characteristic. Such changes can be of many kinds, for example: drug resistance; temperature sensitivity; auxotrophism (requiring a nutrient that the parent does not need for growth); sporulation deficiency; loss of motility; inability to grow anaerobically or aerobically.

Crossing-over between chromosomes

Two chromosomes become associated in a region where their DNA molecules have homology, complete or partial (Fig. 14.1a). The chromosomes are then pictured as intertwining and afterwards separating in such a way that each chromosome acquires a part of the other. Notice that a single crossover leads to changes in two double-stranded molecules of DNA. The process is usually drawn as in Fig. 14.1b, but be aware that neither diagram represents the actual molecular mechanism of the crossover; only the outcome is shown.

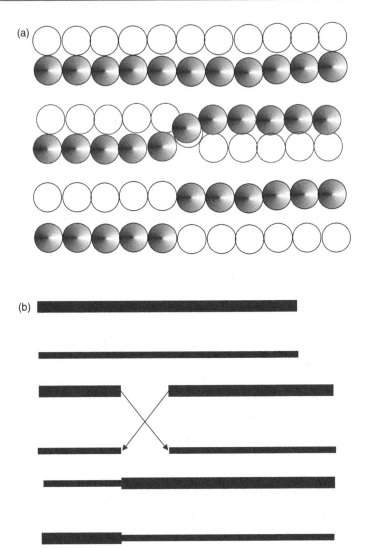

Fig. 14.1 (a) Stages in crossing-over between two chromosomes that associate in homologous regions of their DNA. The two DNAs are double-stranded, but are shown here as single chains for clarity. (b) Another way of visualising the process of crossing-over. The two homologous regions of two molecules of DNA (one shown as thicker in order to distinguish it from the other) are pictured as opening and exchanging material. (c) Insertion of circular DNA (a plasmid carrying a gene for resistance to chloramphenicol) by a single crossover between homologous regions of the plasmid and host DNA. The DNA of the plasmid with the gene for resistance becomes incorporated into the chromosomal DNA of the host.

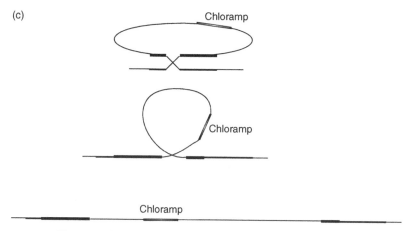

Fig. 14.1 (cont.)

Phage or plasmid DNA may also be integrated into bacterial DNA by a single crossover (Fig. 14.1c). Additional genes thus become a part of the bacterial genome.

14.3   Restriction endonucleases and mapping their sites of action on a plasmid

Restriction endonucleases are enzymes produced by bacteria to degrade foreign (i.e. not their own) DNA. Many such enzymes are known; each recognises a short sequence of bases within a molecule of DNA and cuts the DNA at, or close to, the recognition site. Opening a plasmid with a restriction enzyme that has only one site of action (on that particular plasmid) allows the insertion of DNA to create a cloning or expression vector, which is the basis of recombinant DNA technology. A plasmid breaks into fragments when degraded by a restriction endonuclease that has more than one site of action (Fig. 14.2).

These fragments can be separated by gel electrophoresis, where their mobilities are inversely related to their sizes (measured in kilobase pairs, kbp). Analysis of the sizes of fragments allows the sites of action to be mapped on the plasmid. The next paragraphs illustrate how to do this.

Restriction enzyme 1 yields three fragments from a plasmid: 1.0, 1.5, 2.5 (all kbp). We can use this information to draw Fig. 14.3a. Restriction enzyme 2 yields two fragments: 2.0, 3.0 (both kbp) and so there must be two sites of action (Fig. 14.3b).

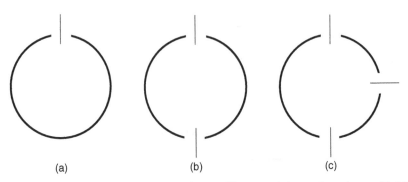

Fig. 14.2 A circular plasmid opened by a restriction endonuclease with (a) one site of action; (b) two sites; (c) three sites. Note that the number of fragments equals the number of sites of action.

To find how the enzyme 2 sites relate to the enzyme 1 sites we look at the fragments produced when the two enzymes act together. Restriction enzyme 1 + enzyme 2 yields five fragments: 0.3, 0.7, 1.0, 1.2, 1.8 (all kbp).

Deducing the map is now an exercise in solving a puzzle in logic. Notice that enzyme 2 breaks up the 2.5 kbp fragment (that is formed by enzyme 1 alone), and also breaks up the 1.5 kbp fragment (from enzyme 1 alone). The 1.8 kbp fragment (seen with both enzymes together) can only come from the 2.5 kbp fragment (enzyme 1 alone) because the two other fragments formed by enzyme 1 alone are already smaller than this. The other enzyme 2 site must be within the 1.5 kbp fragment (enzyme 1 alone). Now we can draw maps for the possible sites of both enzymes (Fig. 14.3c).

The line of argument used here cannot always be applied in exactly the same way to deduce every restriction map. A bit of ingenuity is sometimes required! Let's go a bit further – suppose a third restriction endonuclease (enzyme 3) cuts the above plasmid just once. Where is its site of action? Enzyme 3 + enzyme 1 yields fragments of 0.4, 0.6, 1.5 and 2.5 kbp. Thus it is clear that enzyme 3 cuts within the 1.0 kbp fragment that enzyme 1 alone produces. Enzyme 3 + enzyme 2 yields fragments of 0.7, 1.3 and 3.0 kbp. Now the enzyme 3 site can be located – do you see how – it's easy?

Southern blotting

Genomic DNA can be split into many fragments by one or more restriction endonucleases. Southern blotting allows one to isolate and recognise a single fragment or fragments out of the many generated by the restriction enzymes

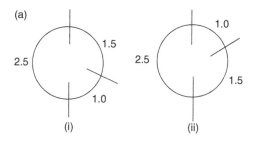

(a)

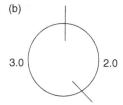

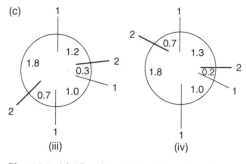

Fig. 14.3 (a) The plasmid is cut into three fragments by enzyme 1, and so there are three sites, separated by the distances shown in kbp. Notice that the two seeming possibilities are really the same (for mapping purposes); if the site at 6 o'clock in (i) is rotated to 12 o'clock then (i) becomes the mirror image of (ii). Drawing the 2.5 kbp fragment on the right (instead of on the left) produces mirror images of (i) and (ii). Putting a site at 12 o'clock is conventional and does not imply that the restriction site is at the origin of replication. (b) Restriction map for endonuclease 2. The site at 12 o'clock is not meant to be coincident with one of the enzyme 1 sites, but is put there by convention. (c) There are two possible sites for enzyme 2 to cut the 2.5 kbp (produced by enzyme 1 alone) fragment, and again there are two possible sites for enzyme 2 to cut the 1.5 kbp (enzyme 1 alone) fragment into 0.3 kpb and 1.2 kbp fragments. Only possibility (ii) shows correctly the sizes of the five fragments found when the two endonucleases act together. Hence, (iii) is the restriction map.

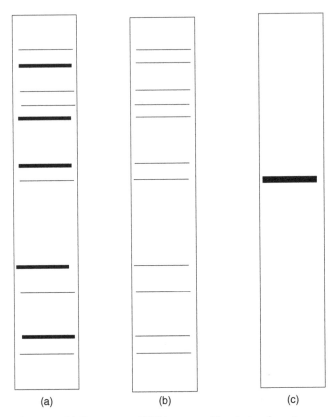

Fig. 14.4 (a) Fragments of DNA separated by electrophoresis on agarose gel. (b) Fragments denatured (to single-stranded DNA) and transferred to a nitrocellulose sheet. (c) Autoradiogram (an X-ray film exposed to the nitrocellulose) after hybridising with a ^{35}S-labelled probe complementary to a particular sequence in the DNA.

that contain specific sequences of bases present in the labelled probe. How this is done is shown in Fig. 14.4. Note that the fragment which is identified may contain many more bases than just those which hybridise with the labelled probe, and that two or more fragments may be detected if the probe sequence contains one or more targets for the restriction enzymes.

Transposons

These are mobile regions of DNA that encode genes needed for their duplication and then the migration of the copy from one region of a chromosome to another region. An insertion sequence is a relatively short transposon (700 to 1600 base pairs) and contains only the genes to enable mobility, flanked by inverted repeat sequences that are 15 to 25 base pairs long. Composite transposons are larger and encode additional genes; frequently these convey resistance to antibiotics. Genes at the new insertion site may be disrupted and mutated by the transposon and some transposons contain transcription promoters or termination codons that may affect the expression of genes near the new point of insertion.

14.4   Here now is a problem that calls for some knowledge of the topics discussed above

The plasmid pAZ1 is an integrational vector for *Bacillus subtilis*. This vector can replicate in *Escherichia coli*, but not in *B. subtilis*, and it carries an antibiotic-resistance gene, conferring chloramphenicol resistance, which is capable of being expressed effectively in both hosts. It can only be maintained in *B. subtilis* if it becomes integrated into the chromosome. The plasmid DNA is circular, 3.3 kbp, and contains a single *Eco*RI site, but no targets for *Hind*III or *Bam*HI.

A 1 kbp *Eco*RI fragment of *B. subtilis* genomic DNA, carrying part of a spore germination gene, has been cloned into the unique *Eco*RI site in vector pAZ1, to yield plasmid pAZ2. This 1 kbp cloned fragment was purified from the plasmid, labelled, and used to probe a Southern blot of chromosomal DNA from wild-type *B. subtilis*, digested with a variety of enzymes. It detected bands at 1.0 kbp in an *Eco*RI digest, 0.6 kbp and 2.0 kbp in a *Hind*III digest, and 0.3 kbp and 0.7 kbp in an *Eco*RI + *Hind*III double digest.

Protoplasts of *B. subtilis* were transformed with pAZ2, and chloramphenicol-resistant transformants selected.

The 1.0 kbp cloned *Eco*RI fragment from pAZ2 was then used to probe Southern blots of *Eco*RI, *Hind*III and *Eco*RI + *Hind*III double digests of DNA from a chloramphenicol-resistant transformant, in order to determine whether integration had occurred by the expected single crossover between the cloned DNA in the plasmid and the homologous sequences in the chromosome.

(1) Draw out the restriction map of this region of the wild-type *B. subtilis* chromosome.

(2) By reference to a restriction map, predict the sizes of hybridising fragments in Southern blots of chromosomal DNA from the *B. subtilis*

transformant, if integration has happened as a result of a single cross-over between the homologous regions on plasmid and chromosome.

(3) List the expected fragment sizes that would hybridise in Southern blots of transformant DNA if vector plasmid pAZ1 were used as probe.

(4) The 1kbp cloned fragment contains the promoter and the first half of the open reading frame of the spore germination gene. Would you expect the B. subtilis transformants to have a wild-type or mutant spore germination phenotype? Explain your answer.

Answers

(1) See Fig. 14.5.

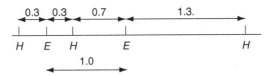

Fig. 14.5 Wild-type chromosome.

(2) See Fig. 14.6.

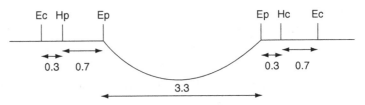

Fig. 14.6 Result of crossover.

EcoRI	1.0 kbp
HindIII	0.6, 2.0, 4.3 kbp
EcoRI + HindIII	0.3, 0.7 kbp

(3)
EcoRI	3.3 kbp
HindIII	4.3 kbp
EcoRI + HindIII	3.3 kbp

(4) Wild-type. There is an intact promoter and part of the gene to the right of the inserted plasmid. This connects directly to the remainder of the gene on the bacterial chromosome.

15 | Problems

Do not be put off by long questions. Often the problem is easier the more information you are given at the start. Often too a long question helps to show the way through an intricate calculation. Short questions can be the hardest: 'What is the formula of water?' for a chemist; 'Is this a question?' for a philosopher.

Several of the problems have been used as examination questions. A few of these you may think entirely unreasonable to face under testing conditions, but remember that candidates would have had previous experience of these difficult topics in practical classes or tutorials.

The problems are more directed to number-crunching than to deduction. Partly this is the author's preference: you can't make valid deductions if you can't do the sums properly. Another reason for the bias is that deduction usually calls for some extra knowledge beyond the information that is given in the question. Not all readers of this book will have such knowledge.

Marking numerical questions isn't always as easy as you may imagine. If the answer is right, then OK, it is easy. Tracing where a wrong answer went wrong (which can take a lot of time) is necessary because the marker does want to give as much credit as possible when the candidate has got nearly everything right, but has made a slip near the end.

As opener, a short problem that takes you right back to the Introduction.

Problem 1

Granules of poly β-hydroxybutyrate are formed in the cytoplasm of many bacteria. After it has been released from the organisms, the polymer can be

151

dissolved in chloroform at 50 °C. In hot concentrated sulphuric acid the dry polymer is quantitatively degraded to crotonic acid:

$$(CH_3-CH-CH_2-CO-)_n \quad \rightarrow \quad nCH_3-CH = CH-COOH$$
$$\underset{-O}{|}$$

poly β-hydroxybutyrate   crotonic acid

Calculate the % (w/w) of poly β-hydroxybutyrate in *Bacillus megaterium* from the following information. The dry wt of the bacteria from 20 ml of a suspension of washed organisms in water was 24 mg. The organisms from a further 5 ml of the same suspension were broken quantitatively by ultra-sonication and the debris (walls and granules) was sedimented by centrifuging. This material was repeatedly extracted with hot chloroform and the total volume of these extracts was made up with chloroform to 50 ml. A sample (1 ml) of this solution was dried in a test tube and sulphuric acid (10 ml) was added to the solid residue. After heating at 100 °C for 10 min and then cooling, the optical density of the resulting solution of crotonic acid in sulphuric acid was 0.438 at 235 nm (1-cm lightpath) measured against a zero of sulphuric acid. Under these conditions, the extinction coefficient of crotonic acid is 1.56×10^4 L mole^{-1} cm^{-1} at 235 nm.

Atomic weights are : C, 12; H, 1; O, 16.

Next, to brush up some maths and algebra, and to jump around connected topics:

Problem 2

Calculations related to logarithms. (**No calculator should be used or needed until question 24.**)

Simplify the following expressions:

(1) $(x^2)^2 / x^3$
(2) $2x^3 / x^2 \times y^2$
(3) $(x^4)^3$
(4) $(3x^3)^3$
(5) $(\sqrt[3]{x})^6$
(6) $x^2 \times \sqrt[4]{x} / x^{3.5}$
(7) $x^2 \times x^{3/2} \times x^{-3}$
(8) $(x^2 + y^2) / y^2$

Find the values of x in the following equations:

(9) $\log_3 81 = x$

(10) $\log_{25} x = 0.5$

(11) $\log_x 49 = 2$

(12) $\log_2 x^3 = 6$

Given that $\log_b 3 = 0.477$ and $\log_b 4 = 0.602$, find:

(13) $\log_b 9$

(14) $\log_b 12$

(15) $\log_b 0.75$

(16) $\log_b \sqrt[3]{12}$

(17) $\log_b (4 \times \sqrt[2]{3})$

(18) $\log_b 6$

Rearrange the following equations to express x in terms of c and y (i.e. as $x = f(c,y)$):

(19) $y = \log_c x$

(20) $y = \log_c x^2$

(21) $y = \log_x c$

(22) $y = \log_c^2 \sqrt{x}$

(23) Find, from the equation $y = c^x$, two ways of expressing x in terms of c and y, and hence derive the relation: $\log_c y = \ln y\, /\, \ln c$

(You need a calculator from now on.)

Find the logarithms to base 2 of the following numbers:

(24) 36

(25) 0.752

Evaluate:

(26) $3.235^{2.2}$

(27) $4.8^{-2/3}$

(28) $\sqrt[3.2]{7}$

(29) $2^{0.25}$

(30) $\sqrt[0.25]{2}$

(31) 3^e

(32) $\sqrt[\pi]{3}$

(33) If you put £100 in the bank and left it to accumulate interest (5% given once annually), how much would be in your account after 10 years? If you had a bacterial culture that contained 100 organisms that were growing exponentially (specific growth rate 0.05 min^{-1}), how many organisms would be present after 10 min? When the different units (£ and organisms) are disregarded, are the two answers numerically the same? Explain your finding.

(34) Radioactive phosphorus (^{32}P) has a half-life of approx. 14 days. If on day 1 you had a sample of $Na_2H^{32}PO_4$ that gave 5000 dpm, what would be its activity (dpm) on day 23? Evaluate the decay constant of ^{32}P.

(35) A bacterial culture was inoculated and samples were taken at intervals during its incubation to measure the optical density (proportional to organisms ml^{-1}), as shown below:

Time of sampling (min after start of incubation)	Optical density
290	0.025
368	0.062
401	0.090
479	0.200
530	0.361
590	0.799
660	1.40
720	1.90
780	2.30
840	2.50
900	2.60

Determine graphically the doubling time of the bacteria (and hence their specific growth rate) during the exponential phase of growth and mark on the graph (as closely as possible) the end of this phase. When do you think the exponential phase began?

(36) A suspension of bacteria was put into a water bath at 60 °C. At intervals, samples were taken and their viable counts (at 30 °C) were found on solid medium after suitable dilution:

Time at 60 °C (min)	Number of viable organisms ml^{-1}
0	1.58×10^9
8	1.41×10^8
15	2.14×10^7
24	1.40×10^6
33	1.26×10^5
48	2.00×10^3

Plot $\log_{10}$ (number of organisms ml^{-1}) against time. From the graph determine:

(a) The decimal reduction time (the time needed at 60 °C to kill 90% of the organisms).

(b) The time needed at 60 °C to kill half the bacterial population.

(c) The percentage of the initial population that survives after 30 min at 60 °C.

(37) What is the pH value of 2M-H_2SO_4?

(38) What is the pH value of 0.3M-NaOH? (The ionic product of water is 1×10^{-14} moles^2 L^{-1}.)

(39) What is the extinction of a solution that has percentage transmittance value of 15%, with 1-cm lightpath? (% Transmittance $(T) = 100 \, I / I_0$; Extinction $= \log_{10} (I_0 / I)$).

(40) What is the percentage transmittance of a solution that has an extinction of 0.5 with 1-cm lightpath?

Problem 3

A wild-type strain of *Escherichia coli* was grown in a minimal medium with glucose (5 g L^{-1}) as source of carbon and energy. The organisms were harvested, washed and suspended in buffer, then assayed for protein and for β-galactosidase activity. The substrate used in this latter assay was ONPG (*o*-nitrophenyl β-D-galactopyranoside), which, when freely accessible to the enzyme, is hydrolysed by β-galactosidase at the same rate as lactose (4-O-β-D-galactopyranosyl-D-glucopyranose). In all the enzymic assays (see below) the relation between time of incubation (of enzyme and substrate) and amount of product formed was linear. No product was formed in any of the controls (ONPG without enzyme; enzyme without ONPG).

The suspension of washed organisms (1 ml) hydrolysed 1.2 μmol of ONPG in 30 min and after dilution (1 ml + 49 ml of water) a sample (0.5 ml) contained 46 μg of protein. The bacteria in part of the undiluted suspension were broken in a French pressure cell, and the assays were repeated with the soluble extract. The extract (1 ml) hydrolysed 0.8 μmol of ONPG in 10 min and after dilution (1 ml + 99 ml of water) a sample (0.5 ml) contained 20 μg of protein.

The same strain was grown in the minimal medium with lactose (5 g L^{-1}) replacing glucose. The washed organisms and an extract were assayed as before.

This second suspension of washed organisms (1 ml) hydrolysed 1.8 μmol of ONPG in 30 min and after dilution (1 ml + 99 ml of water) a sample (1 ml) contained 30 μg of protein. The second extract (0.1 ml) hydrolysed 2.5 μmol of ONPG in 5 min and after dilution (1 ml + 99 ml of water) a sample (0.5 ml) contained 10 μg of protein.

(1) Express the enzymic activity of each suspension and extract as μmol ONPG hydrolysed min^{-1} (mg protein)$^{-1}$.
(2) Interpret the results.

Problem 4

(1) A culture of *Escherichia coli* grew to a final population of 2×10^9 organisms ml^{-1}. The organisms from 1 L of this culture were collected and had a dry weight of 2 g. From a quantitative elementary analysis of these dry bacteria one can estimate that the average weight of a single atom in these organisms is approximately 10 Da.

Suppose that the average molecular weight of all the components of these dried organisms is 1×10^x Da. Hence calculate how many atoms are present in one (dried) organism. (Avogadro's number = 6×10^{23} molecules per mole.)

(2) The internal dimensions of these bacteria are 2 μm × 0.5 μm (assume a cylindrical shape, i.e. flat ends).
(a) Calculate the volume of one organism in m^3.
(b) How many H$^+$ ions are present inside one organism if the internal pH value is 7?

Problem 5

Staphylococci can grow as a biofilm anchored to glass. Calculate the minimum average thickness of one such biofilm (in terms of the number, not necessarily integral, of layers of organisms) from the following information.

A biofilm was grown on the upper surface of a rectangular glass slide (7 × 25 mm) with ^{3}H-thymidine (40 mg L^{-1} ; specific activity 2.75 × 10^6 dpm per mg) in the chemically defined medium, though this nucleoside is not required as a growth factor by staphylococci. The radioactivity of the washed organisms collected from the biofilm was 2.0 × 10^3 dpm. The dry wt of one staphylococcus is 5 × 10^{-13} g and thymidine (after incorporation into DNA) makes up 0.3% of this weight. Determine the total number of organisms on the slide and compare this value with the number of organisms needed to cover the whole surface of the slide as a single close-packed layer (see Fig. 15.1). Assume that each organism is spherical, with a diameter of 1 μm.

What is the fraction (%) of the total organisms in the biofilm that could be directly attached to the glass? Give as many reasons as you can why this result is very likely to be greater than the true fraction that is attached.

The number (n) of circles of radius r that can be close-packed inside a rectangle of sides A and B is given by the equation $n = (A \times B) \div (r^2 \times 2 \times \sqrt{3})$, where A, B and r are all expressed in the same units, and A and B are much larger than r.

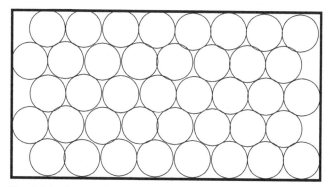

Fig. 15.1 Close-packing of circles into a given area.

Problem 6 (Dr M. M. Attwood)

An aerobic organism is to be grown in mineral medium A. A stock solution
for this medium contains:

K_2HPO_4	34.8 g L^{-1}
$NaH_2PO_4.2H_2O$	31.2 g L^{-1}
$(NH_4)_2SO_4$	40.0 g L^{-1}
$MgSO_4$	4.0 g L^{-1}
$CaCl_2$	200 mg L^{-1}
$FeSO_4$	100 mg L^{-1}
$MnSO_4$	50 mg L^{-1}

50 ml of stock solution and 100 ml of 0.2 M fructose (as source of carbon)
are mixed and then diluted to 1 L with distilled water to make the final
medium A.

(1) Aerobic microorganisms have a growth yield of 0.7 g per g of substrate
when grown on sugars, and the nitrogen content of the bacteria
accounts for 15% of the dry weight. Is this medium carbon or nitrogen
limiting for growth? Give your reasons.

   The organism was grown in this medium and harvested when 10 mM
fructose had been consumed.

(2) Crude extracts of the organism were prepared and assayed for the
enzyme fructose-1,6-bisphosphatase. The assay mixture (1 ml final
volume) contained: Tris buffer, pH 8.0, 50 mM; $MgCl_2$, 5 mM; EDTA,
0.01 mM; NADP, 0.4 mM and linkage enzyme to NADP, 3 units.

   The reaction was started by the addition of fructose-1,6-bisphosphate
to 2 mM final concentration in the assay mixture.

   To make each assay mixture, a stock solution (5 ml) of each reagent
was first prepared. Given that the linkage enzyme contained 1 mg protein
ml^{-1} and contained 300 units mg^{-1} calculate the amount of each reagent
that you would use to make each stock reagent solution and the volume
of each such solution that you would add to an assay mixture.

(3) The crude cell-free extract was purified using an acid precipitation. The
supernatant liquid was then made 50% saturated with ammonium
sulphate and then centrifuged. The supernatant liquid contained the
enzymic activity. Complete the following purification table given that

the protein concentration (mg ml^{-1}) in the individual extracts were as follows: crude extract 1.9; acid supernatant 1.1; and ammonium sulphate supernatant 0.481. The extinction coefficient for NADPH is 6.22×10^3 L mole^{-1} cm^{-1} at 340 nm.

Treatment of extract	Vol. in assay (ml)	Total vol. (ml)	Δ_{340nm} optical density min^{-1}	Activity (nmol min^{-1} ml^{-1})	Total activity (nmol min^{-1})	Specific activity (nmol min^{-1} (mg protein)$^{-1}$)
Crude extract	0.01	25	0.209			
Acid supernate	0.01	27.5	0.121			
(NH$_4$)$_2$SO$_4$ supernate	0.02	30.5	0.151			

Atomic weights

Hydrogen	1
Carbon	12
Nitrogen	14
Oxygen	16
Sodium	23
Magnesium	24
Phosphorus	31
Chlorine	35.5

Molecular formulae

Tris (hydroxymethyl) methylamine NH$_2$.C(CH$_2$OH)$_3$
EDTA (ethylenediaminetetraacetic acid) [CH$_2$N(CH$_2$COOH)$_2$]$_2$
NADP Na$_4$C$_{21}$H$_{26}$N$_7$O$_{17}$P$_3$
Fructose-1,6-bisphosphate trisodium salt C$_6$H$_{11}$O$_{12}$P$_2$Na$_3$.8H$_2$O

Problem 7 (Professor D. J. Kelly)

An obligately anaerobic bacterium was isolated from soil. The organism had the remarkable ability to grow anaerobically using carbon monoxide (CO) as sole source of carbon and energy. A search was made for enzymes that possibly were responsible for the metabolism of carbon monoxide. It was soon found that extracts of the bacterium could oxidise CO to CO_2 provided that a suitable (artificial) electron-acceptor were present. The best results were obtained by using the dye methyl viologen (MV) as the electron-acceptor; the dye is colourless in its oxidised form but has a deep blue colour when reduced:

$$CO + H_2O + 2MV \rightarrow CO_2 + 2H^+ + 2MV^-$$

This provided a convenient assay for the enzyme; the rate of formation of the reduced MV could be followed spectrophotometrically at 600 nm under anaerobic conditions. In a typical assay, 0.1 ml of extract was added to 2.90 ml of a reaction mixture in a cuvette (1-cm lightpath); the mixture contained 1 mM MV dissolved in 100 mM phosphate buffer (pH 7) saturated with CO. The measured rate of change of absorbance at 600 nm was 0.18 min^{-1}.

(1) Given that the extinction coefficient of reduced MV is $7.4 \times 10^3 \text{ L mol}^{-1} \text{ cm}^{-1}$ at 600 nm, and that the extract contained 14.3 mg protein ml^{-1}, calculate the specific activity of the enzyme in units of nmol min^{-1} (mg protein)$^{-1}$.

(2) The enzyme responsible for the above reaction was purified to homogeneity (100-fold purification from the extract) and found to consist of a single polypeptide chain of molecular mass 60 000 Da. It was also green in colour owing to the presence of the metal ion nickel (as Ni^{2+}) in the protein. Chemical analysis showed that 1 mg of the pure enzyme contained 0.97 μg of nickel. How many moles of nickel are there per mole of enzyme?

The atomic weight of nickel is 59 Da.

(3) Calculate the turnover number of the pure enzyme. Assume that the number of active sites on the enzyme is the same as the number of nickel atoms in one molecule of the enzyme.

(4) Determine $\Delta G_0'$ for the oxidation of CO by MV from these electrode potentials:

$$CO + H_2O \rightarrow CO_2 + 2H^+ + 2e^- \quad E_0' = -0.53 \text{ V}$$
$$MV \rightarrow MV + e^- \quad E_0' = -0.44 \text{ V}$$

Problem 8 (Dr M. Wainwright)

The concentrations of sulphur in the form of $S_2O_3^{2-}$ ion ($S_2O_3^{2-}$–S), and in the form of SO_4^{2-} ion (SO_4^{2-}–S) in a soil treated with 1% (w/w) S^0 were determined before studying of the effects of pesticides on S-oxidation.

(1) Given hydrated sodium sulphate ($Na_2SO_4.10H_2O$) and hydrated sodium thiosulphate ($Na_2S_2O_3.5H_2O$) show how you would prepare a series of solutions (for a standard curve) in the range 20 to 100 µg sulphur as $S_2O_3^2$ or sulphur as SO_4^{2-} ml^{-1} (relative atomic masses: Na, 22.9; S, 32; O, 16).

(2) The following data were obtained when 1 ml samples of such standard solutions were assayed:

SO_4^{2-} –S (µg ml^{-1})	OD in assay	$S_2O_3^{2-}$–S (µg ml^{-1})	OD in assay
10	0.030	15	0.060
25	0.075	30	0.125
50	0.160	60	0.250
100	0.325	100	0.415

Four samples of a soil (each 1 g) were separately extracted by shaking with 25 ml of LiCl.2H$_2$O for 15 min. Each slurry was then filtered and 1 ml of filtrate was diluted to 5 ml with water and analysed for SO_4^{2-}–S. The same procedure was repeated in the analysis of $S_2O_3^{2-}$–S except that four samples (each 5 g) of soil were separately extracted with 20 ml of water and 1 ml of each extract was diluted to 5 ml with water. Determine the concentration of SO_4^{2-}–S and $S_2O_3^{2-}$–S in mg g^{-1} soil given that the following OD readings were obtained when 1 ml of each diluted sample was assayed:

SO_4^{2-}–S	$S_2O_3^{2-}$–S
0.275	0.375
0.325	0.350
0.310	0.350
0.290	0.325

Problem 9

Protein that has been bound to a column of ion-exchange material may be eluted by a buffer that contains increasing concentrations of NaCl. These

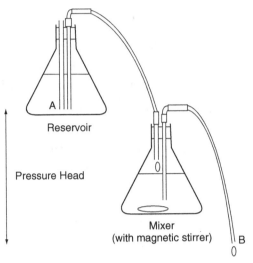

Fig. 15.2 A gradient-forming device. For clarity, tubing is shown with a wider bore than would be used in practice.

increases can be made in steps (e.g. 0.05 M, 0.1 M, 0.15 M, etc. sequentially) or can be made continuously by using a gradient of salt concentration delivered from a mixing device.

A simple type of gradient-former is shown in Fig. 15.2. Two vessels are connected by flexible tubing of narrow bore. (All joints and stoppers must be water-tight!) The reservoir contains buffer with NaCl at a concentration greater than is likely to be needed to elute the protein; the mixer when first attached contains the same buffer but with NaCl at a lower concentration (very often no NaCl). The volume of buffer put in the mixer must be known (see the equation below); the volume in the reservoir is immaterial, provided it is more than enough to complete elution.

If the bottom of the air-inlet tube (A) is higher than the drip-point (B) then brief suction at B will establish a flow out of the mixer and a flow into the mixer (from the reservoir) at exactly the same rate. This rate depends on the pressure head, which is determined by the vertical distance between A and B, and can easily be adjusted by raising or lowering the reservoir.

The concentration of NaCl in the buffer emerging from B at any stage (assuming that the tube from the mixer to B has negligible internal volume) may be calculated from the equation:

$$C_w = C_R - C_D e^{-(V_w / V_m)}$$

where C_w is the concentration being withdrawn when a total volume V_W has emerged from the mixer. C_R is the concentration of NaCl in the reservoir; C_D is the difference in concentrations of NaCl between that in the reservoir and the initial concentration in the mixer; V_m is the volume in the mixer.

A reservoir of 0.5 M NaCl in buffer was connected to a mixer containing 50 ml of the same buffer without NaCl. Calculate the concentrations of NaCl emerging from the mixer when the following volumes (ml) have been withdrawn: 10, 20, 30, 50, 75, 100, 150.

Plot emergent concentration of NaCl (vertical axis) against volume withdrawn (horizontal axis). Is the gradient linear, concave or convex?

The above equation can be rearranged to find at what withdrawn volume a given concentration of NaCl emerges:

$$V_w = V_m \times \ln\left(\frac{C_D}{C_R - C_w}\right)$$

Use this second equation to find the withdrawn volume at which the emergent concentration of NaCl (i.e. C_W) is 0.25 M. V_m, C_R and C_D have the same values as before. Does the answer agree with a read-off from your graph?

Can you derive the second equation from the first? It's a good little exercise in algebra and logarithms.

A partly purified enzyme was loaded onto a column of DEAE-cellulose (interstitial volume 10 ml). The column was first eluted with buffer containing no NaCl, then gradient elution was started and at the same time collection of fractions (each one 5 ml of column effluent) was begun. The reservoir of the gradient-former contained 0.4 M NaCl in the buffer; the mixer (contents 30 ml) initially contained 0.1 M NaCl in the buffer. The desired enzyme was subsequently found only in fraction 7. What does the experiment indicate as the highest concentration of NaCl that did not elute the enzyme, and as the lowest concentration of NaCl that did elute the enzyme?

(Think carefully – an oversight in this kind of situation often leads to a mistake.)

Problem 10 (Dr D. J. Gilmour)

(1) An investigation of the growth of *Bacillus subtilis* in continuous (chemostat) culture was carried out under glucose-limited conditions ($S_r = 5$ mmol L^{-1}). Measurements of the glucose concentration in the culture vessel were made under steady-state conditions, but owing to the closeness of the Christmas holidays only two steady states were achieved. The data are given below.

Dilution rate, D (h^{-1})	Glucose concentration, s (mmol L^{-1})
0.3	0.472
0.6	1.654

Calculate μ_{max} and K_s using the following equation:

$$K_s = \frac{s1(\mu_{max} - D1)}{D1} = \frac{s2(\mu_{max} - D2)}{D2}$$

(2) After the holidays the researcher completed the experiment by measuring s at a full range of steady states. Results are given below.

Dilution rate, D (h^{-1})	Glucose concentration, s (mmol L^{-1})
0.1	0.122
0.2	0.276
0.4	0.734
0.5	1.106
0.7	2.560
0.8	4.405

Using these data, plot $1/D$ against $1/s$ and determine K_s and μ_{max} by this method.

Problem 11 (Professor D. Tempest)

When growing aerobically on glucose in a defined simple medium with ammonia as the sole nitrogen source, *Escherichia coli* expresses a yield value ($Y_{glucose}$) of 100 g mol^{-1}. Elemental analysis of these bacteria (after drying) showed them to contain (by weight) 48% carbon, 32% oxygen, 14% nitrogen and 6% hydrogen.

(1) Write out an empirical formula for the (dried) *E. coli.*

(2) Write out a chemical equation for bacterial growth that shows the stoichiometric relationship between the consumption of glucose, ammonia and oxygen, and the production of biomass, carbon dioxide and water.

(3) Calculate the yield value with respect to oxygen consumption.

(4) Calculate the Y_{ATP} value associated with the growth of *E. coli*, given that 2 molecules of ATP are generated per atom of oxygen consumed in respiration.

Problem 12

A solution (A) of 20 mg of impure adenosine triphosphate (ATP) was made in 10 ml of water. Solution A was diluted (1 ml plus 99 ml of buffer, pH7) to give solution B. The optical density at 259 nm of solution B (1-cm lightpath) was 0.45. Calculate the molar concentration of ATP in solution A, given that the extinction coefficient at pH7 of ATP is 15.4×10^3 L mole^{-1} cm^{-1} at 259 nm. Assume that only ATP itself contributes to the optical density of solution B at 259 nm.

Calculate the purity of the solid ATP as % (w / w), i.e. (wt of pure ATP / wt of impure solid) × 100. The mol. wt of ATP is 507 Da.

What volumes of solution A and of water would you mix to make 50 ml of a solution that was 1 mmolar in respect of ATP?

How would you dilute solution A (state how many ml of solution A you would add to how many ml of water) to prepare solution C, of which a 0.1-ml sample should contain a total weight of phosphorus that falls within the accurate range (1 to 4 μg of phosphorus) of an assay for this element? State the precise concentration of phosphorus (μg ml^{-1}) that you expect to be present in your solution C. Assume that all the phosphorus present in solution A is in the form of ATP. The atomic weight of phosphorus is 31.

Problem 13 (Professor A. Moir)

The plasmid pLX100 carries a gene for ampicillin resistance. Linear DNA fragments of the following sizes were obtained on cleavage of plasmid pLX100 with restriction enzymes:

*Eco*RI digest:	5.4 kbp
*Bam*HI digest:	5.4 kbp
*Pst*I digest:	3.3 kbp, 2.1 kbp
*Eco*RI + *Bam*HI digest:	4.5 kbp, 0.9 kbp
*Eco*RI + *Pst*I digest:	2.3 kbp, 2.1 kbp, 1.0 kbp
*Bam*HI + *Pst*I digest:	2.1 kbp, 1.9 kbp, 1.4 kbp

(1) Deduce the restriction map of the plasmid DNA.

(2) A sample (2 µl) of the plasmid pLX100 DNA (100 µg ml^{-1}) was diluted to a total volume of 1 ml with sterile buffer, and 2 µl of this dilution was used to transform 100 µl of a suspension of competent cells of *Escherichia coli* (ampicillin sensitive). After heat shock, 1 ml of broth was added to the organisms, which then were incubated for 1 h before plating 0.1-ml aliquots on agar containing ampicillin. The average number of colonies obtained per plate was 250; no colonies were obtained on negative control plates.

How many transformants were obtained per 1 µg of pLX100 DNA?

(3) Given that the molecular weight of the average nucleotide in DNA is 330, and assuming that each transformant arose from a separate transformational event, what proportion of the input pLX100 DNA molecules were recovered in transformants? (Avogadro's number is 6.02×10^{23} molecules mole^{-1}.)

Problem 14

A polysaccharide present in walls of *Bacillus megaterium* has the structure:

$$(-\text{Glucose} - \text{Rhamnose} - \text{Rhamnose}-)_n$$
$$|$$
$$\text{Glucuronic acid}$$

After walls have been extensively digested with lysozyme the polymer can be isolated with a fragment of peptidoglycan covalently attached. From its behaviour on gel-filtration the molecular weight of the polymer appears to be >100 000 Da.

Walls were isolated from a mutant (*dap*$^-$ *lys*$^-$) of *B. megaterium* that had been grown in a chemically defined medium containing [^{14}C]diaminopimelate (400 kBq L^{-1}) with an excess of unlabelled diaminopimelate

(100 mg L^{-1}) and L-lysine (100 mg L^{-1}). The presence of the radioactive material did not significantly alter the total concentration of diaminopimelate in the medium.

These walls were digested with lysozyme and the polysaccharide was isolated in aqueous solution. A sample (0.5 ml) of this solution gave 1000 dpm. Another sample (1 ml) of the solution was assayed for rhamnose and found to contain 40 μmol of this deoxyhexose.

The molecular weights of components of the polysaccharide are:

Glucose repeat unit:	162 Da
Rhamnose (monosaccharide) repeat unit:	146 Da
Glucuronic acid repeat unit:	176 Da

One molecule of diaminopimelic acid (Dap; mol. wt 190 Da) is present in each polymer unit of peptidoglycan: N-acetyl-glucosamine-N-acetyl-muramyl-ala-glu-Dap-ala-ala.

Assume that one peptidoglycan unit is covalently linked to one polymer unit of polysaccharide, and hence estimate the mol. wt of the polysaccharide from the above data. Ignore the contribution of peptidoglycan to the weight of the polysaccharide. (1 kBq $\equiv$ 6 $\times$ 10^4 dpm.)

Problem 15

A pure culture of a bacterial species was isolated from pasteurised soil. These organisms were incubated at 30 °C in a liquid chemically defined medium with a low concentration of glucose (0.5 g L^{-1}) as sole source of carbon. When growth reached the late exponential phase a sample (0.1 ml) was taken from this culture and quickly mixed with 100 ml of buffer (20 mM; pH 7) that was at 80 °C and was maintained at this temperature.

After 1 min the suspension at 80 °C was sampled by removing 0.1 ml which was at once mixed with 9.9 ml of buffer at 30 °C. To find the number of organisms that had survived at 80 °C this second suspension was plated (0.1 ml and 0.5 ml samples) onto a rich medium and incubated overnight at 30 °C. At intervals a sample (always 0.1 ml) was taken from the suspension at 80 °C and mixed with a separate 9.9 ml of buffer at 30 °C before plating out as above (see the table below).

Time (min) at 80 °C before transfer to 30 °C	Colonies from 0.1 ml sample of buffer suspension at 30 °C	Colonies from 0.5 ml sample of buffer suspension at 30 °C
1	240	Too many for accurate count
2	79	407
3	25	122
4	7	32
6	4	15
8	3	20
12	5	31
18	4	30
30	6	29

(1) Plot $\log_{10}$ (viable count per ml of the suspension kept at 80 °C) against time at 80 °C.
(2) Interpret the graph as fully as you can.
(3) Estimate from the graph the viable count per ml of the culture at 30 °C at the time when it was sampled.
(4) What percentage of the colony-formers that were in the culture at 30 °C (at the time when it was sampled) survive for 30 min at 80 °C?

Problem 16 (Dr M. Wainwright)

For counts of thiobacilli, soil (1 g) was shaken for 15 min in 10 ml of ¼ strength Ringer's solution and five 0.1 ml aliquots were spread (after dilution 10^3-fold) on the surface of thiosulphate medium. The following counts were obtained for two separate soils:

Soil A (colonies per plate)	Soil B (colonies per plate)
10	8
9	8
6	3
9	3
8	8

(1) Determine the average number of thiobacilli per gram of each soil.

(2) Find the median and mean of colonies per plate and the sample standard deviation for soil A and then for soil B. Use both the t-test and Mann–Whitney test to determine whether the mean A is bigger than mean B with $\geq 95\%$ probability.

Problem 17 (Professor J. R. Quayle)

A portion of a culture of *Escherichia coli* growing on acetate (as sole source of carbon) was harvested and assayed for the activity of isocitrate lyase by measurement of the isocitrate-dependent formation of glyoxylate catalysed by an extract of these organisms. The glyoxylate was trapped as glyoxylate phenylhydrazone ($E_{324} = 1.7 \times 10^4$ L mol^{-1} cm^{-1}). One portion of the extract (A) was diluted 10-fold and 0.05 ml of the diluted extract was found to contain 62 µg of protein. A second portion of extract A was diluted 50-fold and used for the spectrophotometric estimation of isocitrate lyase activity. In this determination diluted extract (75 µl) catalysed a linear rate of change of absorbance at 324 nm of 0.863 in 5 min; the final volume in the spectrophotometer cuvette was 3 ml.

The carbon substrate in the original culture was then changed from acetate to succinate, and after a short lag phase the organisms grew exponentially on the new substrate for 18.5 h. What would you expect the final specific activity (nmol min^{-1} (mg protein)$^{-1}$) of the isocitrate lyase to be if the mean generation time of the organisms with succinate was 3.9 h?

During growth on acetate as sole carbon source the first step in the entry of acetate into cellular metabolism is its activation to acetyl CoA via acetate thiokinase. Half of the resulting acetyl CoA is totally oxidised to carbon dioxide to provide energy, and the other half is made into cellular material. Fixation of carbon dioxide is negligible. If the activity of acetate thiokinase were the rate-limiting step for growth on acetate, what maximum rate of exponential growth (expressed in terms of mean generation time) could this enzyme support if its specific activity were 0.2 µmol acetate activated min^{-1} (mg protein)$^{-1}$? Assume that 50% of the dry weight of an organism is carbon and 50% is protein.

To answer this question you need to know that *E. coli* makes isocitrate lyase (so that the glyoxylate cycle can operate) when growing on acetate, but the formation of this enzyme stops when an intermediate of the citric acid cycle (such as succinate) is available. When the acetate-grown organisms are shifted to a medium containing succinate the isocitrate lyase that had already been formed is not destroyed, but the molecules of this enzyme become distributed among an increasing number of bacteria as the culture grows on succinate. Consequently the specific activity of isocitrate lyase will fall.

Problem 18 (Professor J. R. Quayle)

A culture of photosynthetic bacteria was grown anaerobically in the light. A sample was taken from the culture, and diluted (0.5 ml + 3.5 ml water) before measuring its optical density, which was 0.31. At the same time of sampling, a further 100 ml was removed from the culture and the organisms were centrifuged, then bacteriochlorophyll was extracted from them with ether. The total volume of the ether extract was 250 ml, and its absorbance at 740 nm was 0.50.

Calculate moles of chlorophyll per g bacteria, given that an optical density of 1.0 was given by a suspension containing 1.134 mg dry wt bacteria ml^{-1}, and that a 1% (w/v) solution of bacteriochlorophyll in ether has an absorbance at 740 nm of 1.054×10^3. The mol. wt of bacteriochlorophyll is 911.

The culture was switched to aerobic growth in the dark. Formation of bacteriochlorophyll stopped at once, but exponential growth continued with a doubling time of 5 h. If there is no destruction of bacteriochlorophyll, what will be the content (moles of chlorophyll g^{-1} bacteria) after the first 12 h of exponential growth?

Problem 19 (Professor J. R. Guest)

The products of digestion of DNA with restriction endonucleases can be separated by electrophoresis in agarose gel, and can be detected by their fluorescence in UV light after treating with ethidium bromide. The relationship between DNA size (in kbp) and electrophoretic mobility can be shown by plotting $\log_{10}$ size against distance run (mm) by standards.

Size of standard (kbp)	Distance moved by standard (mm)
50	21
22	27
14	4
9.4	45
6.8	61
5.3	75
4.2	90
3.1	110
2.3	134
1.9	150
1.3	180

A segment of *Escherichia coli* DNA was cloned into a λ-vector phage to produce a lysogenic transducing phage that could restore the Mic$^+$ phenotype to *micA* mutants. Samples of DNA from the transducing phage and from the vector phage were digested with two endonucleases: *Hind*III and *Eco*RI. Mobilities of the fragments of digestion on agarose are shown below.

Distances moved (mm) by fragments obtained with:

*Hind*III		*Eco*RI		*Hind*III + *Eco*RI	
λ-vector	λ-*mic*	λ-vector	λ-*mic*	λ-vector	λ-*mic*
26	26	27	27	27	27
32	32	40	36	47	47
	78	59	59	59	59
			106	150	99
					150
					185

(1) Plot the standard curve, and read off the sizes (in kbp) of the fragments of digestion.

(2) Derive restriction maps for the vector and for the transducing phages, assuming that the large *Hind*III fragment (mobility 26 mm) corresponds to the left end of the *linear* phage maps.

(3) State the overall sizes of the vector and transducing phages in kbp, and the size of the bacterial fragment cloned.

Problem 20

A bacterium excreted lysine after a batch culture (in minimal glucose + salts medium) had reached the stationary phase. In an experiment, growth was assessed from the turbidity of the culture, and lysine was measured in samples (0.5 ml) of the medium, by a specific assay with ninhydrin, in which the optical density at 420 nm was read.

Time (h) of incubation	Turbidity of culture (colorimeter reading)	Lysine assay (extinction at 420 nm)
9	0.05	0
11	0.1	0
13	0.3	0
15	0.5	0
17	1.0	0
20	2.3	0
21	2.8	0.04
23	2.7	0.28
25	2.7	0.52
27	2.8	0.73
29	2.8	0.96

The relation between turbidity and dry wt of bacteria was linear up to a colorimeter reading of 3.0, and a reading of 1.0 corresponded to a dry wt of 0.35 mg ml^{-1}. The relation between optical density at 420 nm and the amount of lysine present in the assay system was linear up to an optical density of 1.5, which corresponded to 1.8 µmoles of lysine.

Plot $\log_{10}$ (turbidity of the culture × 100) against time of incubation, and hence determine the doubling time of the organisms during the exponential phase of growth.

On the same graph plot the concentration of lysine in the medium against time of incubation, and hence determine the rate of lysine production. Express this rate as nmoles min^{-1} (mg dry wt bacteria)$^{-1}$.

A new assay for one of the enzymes of lysine biosynthesis (tetrahydrodipicolinate acetylase) was devised. In this assay, an extract of stationary phase organisms (harvested after incubation for 25 h) showed an activity of only 10 nmoles (acetyl coenzyme A used) min^{-1} (mg protein)$^{-1}$.

(1) Is this rate sufficient to account for the observed rate of lysine excretion? Assume that 50% of the bacterial dry wt is protein.

(2) Can the enzymic activity account for the necessary rate of lysine formation (to make protein and cell wall) during exponential growth? Assume that 5% of the bacterial dry wt is lysine, and that diaminopimelate is not used structurally. The specific growth rate (g of new organisms formed per g of existing bacteria per unit time) is given by (ln 2) ÷ (doubling time); and the mol. wt of lysine is 146.

Problem 21 (Professor A. Moir)

A 2.5 kbp *Eco*RI fragment of *Bacillus subtilis* DNA was cloned in the plasmid vector pAB1 to yield hybrid plasmid pAB2. The vector, pAB1, is a 4.2 kbp plasmid, with a multiple cloning site containing targets for three enzymes, in the order *Bam*HI – *Eco*RI – *Hind*III. These are the only target sites for these enzymes in plasmid pAB1.

Restriction digests of the plasmid pAB2 gave fragments (kbp) as follows:

*Eco*RI	4.2, 2.5
*Hind*III	4.7, 1.3, 0.7
*Hind*III + *Eco*RI	4.2, 1.3, 0.7, 0.5
*Bam*HI	5.9, 0.8
*Bam*HI + *Hind*III	4.2, 1.3, 0.5, 0.4, 0.3

(1) Deduce the restriction map of the plasmid pAB2, explaining your conclusion.

(2) A culture of cells containing the plasmid pAB2 was grown. Samples were removed for viable count, serially diluted and 0.1ml of a 10^{-5} dilution was plated, giving replicates of 210 and 190 colonies. Plasmid DNA was isolated from 100 ml of the culture, and yielded 20 μg. Assuming 100% recovery of plasmid DNA during the isolation procedure, what is the approximate copy number of the plasmid pAB2 per cell? (One nucleotide in DNA has an average molecular weight of 300. Avogadro's number is 6.02×10^{23}.)

Problem 22

(1) Pure DNA was isolated from bacteria of strain X. The four heterocyclic bases in this DNA were released quantitatively by hydrolysis with acid.

Paper chromatography was used to separate these bases, the positions of which were located by examination of the dried chromatogram under a UV lamp. An area of paper containing the thymine spot was cut out and the base was eluted quantitatively into 5 ml of 0.1 M HCl. Three more solutions of the other bases from the chromatogram were similarly prepared. Use the data below to deduce the %GC of this DNA.

	Absorbance (1-cm lightpath) of the solution in 0.1 M HCl at wavelength of maximum absorbance	Millimolar extinction coefficient of the base (in 0.1 M HCl) at wavelength of maximum absorbance (L mmol^{-1} cm^{-1})
Thymine	0.28	8.0
Cytosine	0.15	10.0
Adenine	0.44	12.6
Guanine	0.17	11.1

(2) What weight (mg) of thymine (mol. wt 126) is needed to synthesise 1 g of this DNA? (The average mol. wt of a nucleotide from any DNA is 307.)

(3) In strain X, DNA represents 5% of the dry weight of the organisms. What, therefore, is the minimum weight (mg) of thymine needed to yield a crop of 1 g (dry wt) of these bacteria? Assume that thymine has no other metabolic role than contributing to the structure of DNA.

(4) Pure radioactive DNA, of specific activity 1×10^6 cpm mg^{-1}, was required from strain X. To label the DNA, you may grow the organisms with either [^{3}H]-thymine or [^{14}C]-thymine in the medium. Efficiencies of counting are: ^{3}H, 20% and ^{14}C, 80%. Unlabelled thymine (10 µg ml^{-1}) is already present, and 1 ml of medium produces 1 mg (dry wt) of bacteria. Addition of radioactive material does not significantly alter the concentration of thymine in the medium. Synthesis of thymine from its precursors by the organisms is totally inhibited when this pyrimidine is present in the medium.

Calculate how much [^{3}H]-thymine (µCi) you would add per litre of medium to get DNA of the required specific activity. How much [^{14}C]-thymine would be needed per litre if this material were used instead of [^{3}H]-thymine to produce DNA of the same specific activity? (1 µCi is equivalent to 2.2×10^6 dpm.)

(5) The smallest package of [^{3}H]-thymine that can be bought costs £137* and contains 1 mCi. The smallest package of [^{14}C]-thymine costs £56* and contains 50 μCi; a package containing 250 μCi costs £183*. The procedure for isolating DNA yields only one-tenth of the total bacterial DNA in pure form. What would you buy in order to prepare as cheaply as possible 10 mg of pure labelled DNA of the desired specific activity? Assume that unused labelled thymine or a surplus of radioactive DNA have no value. (*Price includes VAT at 17.5% and delivery at £10 per package.)

(6) Is the %GC of the DNA relevant to the cost of isolating pure radioactive DNA? What if the %GC had proved to be 25% or 75%?

Problem 23 (Dr M. M. Attwood)

To measure the extent of purification of an enzyme, its activity (μmoles min^{-1} (mg protein)$^{-1}$) was measured in the initial crude cell-free extract and in the final enzymic preparation.

The initial crude cell-free extract was diluted (0.5 ml with 9.5 ml of phosphate buffer pH 7.0) and 10 μl of this diluted extract was used to measure the enzymic activity spectrophotometrically. The change in absorption at 340 nm was measured before and after the addition of the enzyme substrate. The assay volume was 3.0 ml, and 10 μl of the diluted extract showed an initial linear change of absorption of 0.42 units over 7 min, then after the addition of the substrate a linear absorption change of 0.6 units over 3 min.

After a number of purification steps the final enzymic preparation was used to measure the final activity. This final preparation (0.1 ml) was made up to 1.0 ml by the addition of ice-cold phosphate buffer pH 7.0, then 30 μl of this diluted preparation was used to measure the absorbance change at 340 nm. Before the substrate was added the linear rate of absorbance change was 0.12 units over 6 min and after the addition of substrate it was 0.45 units over 2 min. The initial diluted cell-free extract (10 μl) contained 20 μg of protein and the diluted purified extract (10 μl) contained 1.5 μg of protein.

The extinction coefficient at 340 nm = 6.22×10^3 L mole^{-1}cm^{-1}.

Calculate the activity of the enzyme in:

(1) the initial crude cell-free extract
(2) the final purified preparation.

(3) By comparing the two activities, report the factor (i.e. *x*-fold) by which the enzyme had been purified.

Problem 24

(1) When *Staphylococcus* sp. is grown as a batch culture in a chemically defined medium with limiting lysine the compound UDP-*N*-acetylmuramyl-L-alanyl-D-glutamate accumulates in the cytoplasm of the bacteria. This dipeptide was extracted with trichloroacetic acid and then isolated in aqueous solution.

Several tests were made to establish the authenticity of the dipeptide and to determine its concentration in solution.

> The solution was colourless, and after dilution (1 / 100, at pH 7) its absorption spectrum in the range 230–300 nm showed a broad peak at 260 nm. At this wavelength the absorbency was 0.53 and at 280 nm it was 0.25. The extinction coefficient of uridinediphosphate at 262 nm is 1×10^4 L mol^{-1} cm^{-1}, and the ratio of the absorbencies at 280 nm and 260 nm is 0.39 for this compound at pH 7.
>
> Total phosphorus was measured in a sample of the (undiluted) solution: 10 μl contained 0.11 μmol P.
>
> A sample (0.2 ml) of the undiluted solution was hydrolysed (4 M HCl; 4 h at 100 °C) then neutralised and made up to 1 ml with water. A sample (0.5 ml) of this hydrolysate contained 0.44 μmol hexosamine. Another sample (0.2 ml) of the hydrolysate contained 0.21 μmol acetate.
>
> After more drastic acid hydrolysis (6 M HCl; 18 h at 100 °C) thin-layer chromatography of the neutralised hydrolysate followed by ninhydrin treatment showed only two spots, of similar intensity, in the positions of alanine and glutamic acid.

Are these data consistent with the structure of the dipeptide? If so, then what is the concentration of the dipeptide in the aqueous solution?

(2) The rate of the enzymic reaction:

UDP-*N*-acetylmuramyl-L-alanyl-D-glutamate + *meso*-diaminopimelate + ATP →

UDP-*N*-acetylmuramyl-L-alanyl-D-glutamate-*meso*-diaminopimelate + ADP + P$_i$

can be measured from the incorporation of [^{3}H]-*meso*-diaminopimelate into the tripeptide.

The complete assay system contains (all volumes in µl):

5 mM UDP-N-acetylmuramyl-L-alanine-D-glutamate (dipeptide)	8
5 mM [^{3}H]-*meso*-diaminopimelate ([^{3}H]-Dap; 1×10^6 dpm ml^{-1})	4
50 mM ATP	4
0.5 M Tris / HCl buffer, pH 8.4	4
0.2 M MgCl$_2$	2
Enzyme (crude extract; 5 mg protein ml^{-1})	4
Water to	50

All components except the enzyme are brought to 37 °C; then at time 0 the enzyme (also at 37 °C) is added. The reaction is stopped after various times of incubation (see below) and a sample (20 µl) from the assay system is spotted onto the origin line of a paper chromatogram. This is developed with a solvent that causes unbound diaminopimelate to migrate away from the origin, whereas diaminopimelate that is bound as a tripeptide remains at or very near the origin line. After drying the paper at the origin the area is cut out and is counted.

Eight tubes are prepared:

	Time (min)	dpm from origin of chromatogram
Complete system	0	45
Complete system	5	140
Complete system	10	243
Complete system	15	366
System minus dipeptide	15	36
System minus enzyme	15	28
System minus ATP	15	67
System minus [^{3}H]-Dap	15	14

Use these data to express the rate of the enzymic reaction as nmol Dap incorporated min^{-1} (mg protein)$^{-1}$ and then as units of Dap-adding enzyme (mg protein)$^{-1}$.

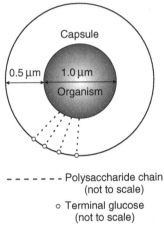

```
- - - - - - - Polysaccharide chain
              (not to scale)
        o  Terminal glucose
           (not to scale)
```

Fig. 15.3 Diagram of organism, capsular polysaccharide and terminal glucose residues.

Problem 25

A single staphylococcus (diameter 1 μm) is surrounded (see Fig. 15.3) by a polysaccharide capsule (uniform thickness 0.5 μm). Assume that half the volume of the capsule is occupied by interstitial molecules of water. Calculate the weight of capsular polysaccharide (the relative density of which is 1.00 after drying) associated with one organism. (The volume of a sphere is given by $4\pi r^3 / 3$, where $\pi = 3.142$ and r is the radius of the sphere.)

The chains of polysaccharide (average mol. wt 500 000 Da) project outwards perpendicularly from the organism (see Fig. 15.3), and each chain has glucose at its terminus (remote from the organism). Calculate how many of these terminal molecules of glucose are exposed to the medium at the outer surface of the capsule of one organism. (Avogadro's number is 6×10^{23} molecules mol^{-1}.)

Calculate the outer surface area of the capsule of one organism and hence determine the number of terminal glucose molecules per μm^2 of capsular surface. (The surface area of a sphere is given by $4\pi r^2$.)

Problem 26 (Professor A. Moir)

A mutant derivative of *Staphylococcus aureus* has been isolated that contains a transposon (Tn*200*) (see Fig. 15.4) in a gene important for virulence (*virA*). This mutant strain was called SA43.

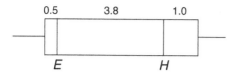

Fig. 15.4 Transposon Tn *200* (5.3 kbp). *E* is an *Eco*RI site; *H* is a *Hin*dIII site.

Chromosomal DNA from this mutant has been isolated, and then digested with restriction enzymes *Eco*RI and *Hin*dIII, singly and in combination. After agarose gel electrophoresis, the digests were probed in a Southern blot.

The probe used consisted of labelled DNA derived from a plasmid carrying the transposon, but carrying no *S. aureus* chromosomal DNA. This probe did not detect any DNA fragments in digests of chromosomal DNA of *S. aureus* that did not contain the transposon.

The results obtained by probing digests of SA43 DNA are tabulated below.

Enzyme(s) used in digest	Size (kbp) of fragments hybridising to probe
*Eco*RI	5.8, 2.3
*Hin*dIII	5.2, 2.5
*Eco*RI + *Hin*dIII	3.8, 2.0, 1.4

(1) Deduce the restriction map of the region of the chromosome of SA43 around and including the transposon insertion, pointing out those fragments that hybridise with the probe in each digest. Outline the stages of your deduction in your answer.

(2) Deduce the restriction map of the chromosome of the wild-type strain in this region.

(3) How many hybridising bands might you have expected to see in a Southern blot of similar digests if a strain were to contain *two* Tn*200* insertions, located in different regions of the chromosome?

Problem 27

Autoclaving an aqueous suspension of bacterial endospores releases quantitatively their content of dipicolinic acid (DPA). The absorption at 440 nm of the Fe^{2+} complex of dipicolinic acid provides a means of assaying this acid in solution. Use the data below to draw a standard curve for this colorimetric assay.

Every tube in the assay contained 4 ml of a buffered solution of ferrous sulphate plus a reducing agent (ascorbic acid). Additions were made as below to give in each tube a total volume of 5 ml before measuring the extinctions against a zero of buffered ferrous sulphate plus ascorbate.

	Tube number					
	1	2	3	4	5	6
Water	1.0 ml	0.9 ml	0.8 ml	0.7 ml	0.5 ml	0
Dipicolinic acid ($1 \, mg \, ml^{-1}$)	0	0.1 ml	0.2 ml	0.3 ml	0.5 ml	1.0 ml
Extinction at 440 nm (1-cm lightpath)	0.002	0.115	0.234	0.361	0.601	0.892

(1) A suspension of spores (50 mg dry wt in 5 ml water) was autoclaved, cooled and the insoluble material removed by centrifuging. Under the standard conditions for assay of dipicolinate a sample (1 ml) of this extract gave an extinction of 0.533. Express the content of calcium dipicolinate in these spores as a percentage (w / w) of the dry wt of the spores. (The mol. wt of dipicolinic acid is 167 Da, and the atomic wt of Ca^{2+} is 40 Da.)

(2) Determine the molar extinction coefficient of the Fe^{2+} complex of dipicolinate at 440 nm.

(3) Dipicolinate in these same spores was assayed by a different method. Dipicolinate was extracted from the spores (2 mg) by heating with 95% (v / v) aqueous ethanol + water + acetic acid (3 : 1 : 0.04 (v / v); 2 ml total vol.) at 100 °C for 1 h in a sealed tube. After centrifuging and removing the supernatant liquid to a clean tube, the solid residue was re-extracted with 80% (v / v) aqueous ethanol (2 ml) then recentrifuged, and the supernatant liquid added to the first extract. This combined extract was evaporated to dryness then 4 ml of 5 mM $Ca(OH)_2$ was added and the mixture again centrifuged. The optical density at 270 nm of the final supernatant liquid (a measure of calcium dipicolinate) was too high to read accurately. After dilution of this liquid (1.0 ml + 4.0 ml water) the

optical density (1-cm lightpath) at 270 nm was 0.154. The extinction coefficient of calcium dipicolinate is 5.4×10^3 L mole^{-1} cm^{-1} at 270 nm. Calculate the content of this compound in the spores as % (w/w). Do the results of the two methods of assay agree?

Problem 28 (Dr D. J. Gilmour)

A thermophilic bacterium was grown in a chemostat culture at 55 °C in a defined medium in which glucose was the limiting nutrient ($S_r = 10$ mmoles L^{-1}). A range of steady states was achieved and at each steady state the biomass, the extracellular concentration of glucose in the culture vessel and the respiration rate (q_{oxygen}) were determined. The results are shown below.

Dilution rate (D) (h^{-1})	Biomass (g L^{-1})	Glucose (μg ml^{-1})	Respiration rate (q_{oxygen}) (mmoles O$_2$ (g bacteria)$^{-1}$ h^{-1})
0.1	2.54	1.7	3.3
0.2	2.46	2.3	4.9
0.3	2.32	3.8	6.5
0.4	2.04	5.5	8.1
0.5	1.74	7.4	9.8
0.6	1.42	10.2	11.0

(1) Plot q_{oxygen} versus μ to determine the maintenance energy (q_m, mmol g^{-1} h^{-1}) with respect to oxygen. Then calculate the yield value attributed to growth alone (Y_g g mol^{-1}) for oxygen.

(2) Plot $1/\mu$ against $1/s$ and hence calculate the maximum growth rate (μ_{max}, h^{-1}) and the saturation constant (K_s, mmol L^{-1}) with respect to glucose.

(3) The experiment was repeated at 70 °C, but owing to problems with maintaining this high temperature only two steady states were achieved and no respiration rates were measured. The results are shown below.

Dilution rate (D) (h^{-1})	Biomass (g L^{-1})	Glucose (μg ml^{-1})
0.3	3.7	2.2
0.4	2.4	5.1

Use the equation shown below to calculate the μ_{max} (h^{-1}) and K_s ($mmol\,l^{-1}$) values at 70 °C.

$$K_s = s_1(\mu_{max} - D_1) / D_1 = s_2(\mu_{max} - D_2) / D_2$$

(The molecular weight of glucose is 180.)

Problem 29

The enzyme transketolase catalyses the formation of erythrose 4-phosphate, which is needed in the formation of aromatic amino acids: one molecule of erythrose 4-phosphate is used in making one molecule of any aromatic amino acid.

(1) Calculate the amount of erythrose 4-phosphate (nmoles) needed to form 1 mg (dry wt) of bacteria. Assume that aromatic amino acids (average mol. wt 165) represent 15% (w/w) of bacterial protein, and that protein makes up 50% of the dry wt of a bacterium.

(2) An extract was made from bacteria out of an aerobic culture that was growing exponentially with a doubling time of 120 min in a medium containing only glucose and inorganic salts. The specific activity of transketolase in this extract was 10 nmol min^{-1} (mg protein)$^{-1}$. Is the rate of growth of these organisms likely to be limited by the activity of transketolase if the action of this enzyme is the only way by which erythrose 4-phosphate can be produced?

Problem 30 (Professor A. Moir)

Plasmid vector pX1, 3.5 kbp in size, can replicate in *Escherichia coli*, but not in *Staphylococcus aureus*. If a fragment of *S. aureus* DNA is cloned into this plasmid, it provides a region of homology that can recombine with the equivalent region on the bacterial chromosome, so that a single crossover results in addition of the hybrid plasmid into the staphylococcal chromosome.

In order to inactivate the *purA* gene, plasmid pX1 was linearised by digestion with *Hind*III and *Eco*RI (these sites are adjacent within a cluster of restriction enzyme sites in the multiple cloning site of the vector) and ligated with a 0.4 kbp *Hind*III–*Eco*RI fragment of *S. aureus* chromosomal DNA. This DNA had been derived from a clone carrying the *purA* gene.

The resulting plasmid pX2 was introduced into *S. aureus*, and recombinants were selected in which this plasmid had integrated into the chromosome. DNA was isolated from one such transformant, digested, and the fragments separated by agarose gel electrophoresis.

A probe was prepared from the entire pX1 vector DNA. No bands were seen when this probe was hybridised to blots of wild-type chromosomal DNA. A probe was also prepared from the 0.4 kbp insert fragment of pX2.

The results of the Southern blots of digests of chromosomal DNA from the transformant are shown below.

Enzyme(s) used in digest	Probe	Size (kbp) of hybridising fragment
*Eco*RI	pX1	3.9
*Hind*III	pX1	3.9
*Eco*RI + *Hind*III	pX1	3.5
*Eco*RI	0.4 kbp insert	3.9, 1.6
*Hind*III	0.4 kbp insert	7.5, 3.9

(1) Draw a diagram of the hybrid plasmid, and the expected integration process.
(2) Draw a restriction map of the region around and including the inserted plasmid, showing clearly how the hybridisation pattern shown above is explained by your proposed map.
(3) For each of three digests of wild-type chromosomal DNA (*Eco*RI, *Hind*III and a double *Eco*RI + *Hind*III digest), predict the size of labelled bands detected if the 0.4 kbp insert were used as probe.

Problem 31

(1) Calculate the energy (in kcal) delivered by one mole of photons of wavelength 740 nm. Determine the potential to which an electron could be driven, by this energy, from the reaction centre (RC) of a green sulphur bacterium, where:

$$RC \rightarrow RC^+ + e^-, \quad E_0' = +0.6\,V$$

(2) Do similar calculations for light of 850 nm wavelength received by the RC of a purple sulphur bacterium, where the RC has an electrode potential of $E'_0 = +0.8$ V.

(3) One candela of white light delivers approximately 2×10^{14} quanta s^{-1} / cm^{-2}. Assume that 10% of these quanta are at wavelengths that are absorbed by the photosynthetic apparatus and that 6.02×10^{23} of such quanta deliver 40 kcal.

How much energy is available for photosynthesis from an illumination of 500 candela of white light falling evenly on a culture of cyanobacteria (100 cm^2 transparent surface area) for 1 h?

Problem 32 (Professor D. J. Kelly)

A sample of [^{35}S]-methionine containing 0.1 μg of this amino acid was counted at 82% efficiency in a liquid scintillation counter, giving a figure of 3570 cpm. The background count was 18 cpm.

(1) Calculate the specific activity of the sample as kBq mg^{-1}.

(2) Another sample from the same batch of [^{35}S]-methionine was used 10 days later to label a protein during an in vitro translation. What is the specific activity of sample (expressed as kBq mg^{-1}) at the time of doing the experiment?

(3) In the labelling experiment, 1 mg of the [^{35}S]-methionine was added to unlabelled methionine in the 1 ml translation mix to give a total concentration of 50 mM methionine. The purified translated protein (0.3 mg) gave a count of 537 dps (corrected for background and counting efficiency). How many μmoles of methionine were incorporated per mg of translated protein?

Given that the molecular weight of the protein is 14 925, how many methionine residues does it contain?

1 Bq = 1 dps (disintegration per second)
cpm = counts per minute
The half-life of ^{35}S = 87.1 days

Problem 33

The enzyme diaminopimelate epimerase catalyses the reaction:

LL-diaminopimelate ↔ *meso*-diaminopimelate

The epimerase gives a mixture of the two isomers, and the optimum pH value is 8.0: there is no activity at pH 10.5. An activator (1 mM 2,3-dimercaptopropan-1-ol) is required. A unit of this enzyme isomerises 1 μmol of LL-diaminopimelate min^{-1} at pH 8 and at 30 °C, when the initial concentration of LL-diaminopimelate is 10 mM.

The enzyme *meso*-diaminopimelate dehydrogenase catalyses the reaction:

$$\textit{meso-}\text{diaminopimelate} + \text{NADP}^+ \rightarrow$$
$$\text{L-}\Delta^1\text{-tetrahydrodipicolinate} + \text{NADPH} + \text{H}^+ + \text{NH}_3$$

The dehydrogenase is specific for the *meso*-isomer, and the optimum pH value is 10.5, though activity is considerable at pH 8.0. One unit of this enzyme catalyses the formation of 1 μmol of NADPH min^{-1} at 30 °C and at pH 10.5, when the initial concentration of *meso*-diaminopimelate is 3 mM.

In order to devise a spectrophotometric assay for the epimerase a series of mixtures was prepared in quartz cuvettes of 1-cm lightpath:

Mixture number	1	2	3	4	5	6	7	8
meso-diaminopimelate (30 mM)	0.3	–	0.3	–	–	–	–	–
LL-diaminopimelate (100 mM)	–	0.3	–	0.3	0.3	0.3	0.3	0.3
Diaminopimelate dehydrogenase (1 unit ml^{-1})	0.5	0.5	–	–	0.5	0.5	0.5	0.5
Diaminopimelate epimerase ((8 mg protein) ml^{-1})	–	–	0.2	0.2	0.05	0.1	0.15	0.2

All quantities in this table are in ml. Every mixture also contained: phosphate buffer, pH 8.0 (300 μmoles); 2,3-dimercaptopropan-1-ol (3 μmoles); NADP (3 μmoles); and water (to 3 ml final volume in every case).

Each mixture was prepared without diaminopimelate, and was equilibrated at 30 °C in a recording spectrophotometer. The appropriate isomer of diaminopimelate was then added, and the initial rate of change of extinction was measured at 340 nm for each mixture in turn, against a zero cuvette which contained all the reactants (as above) except the enzymes. These rates (expressed as change of extinction per min) were:

Mixture 1	0.35	Mixture 5	0.041
Mixture 2	0.00	Mixture 6	0.083
Mixture 3	0.00	Mixture 7	0.11
Mixture 4	0.00	Mixture 8	0.13

(1) What is the ratio of the activity of diaminopimelate dehydrogenase at pH 8.0 to the activity of this enzyme at pH 10.5, with *meso*-diaminopimelate as substrate?

(2) Plot the measured activity (units ml^{-1}) of the epimerase with LL-diaminopimelate against the volume of this enzyme used in the assay. What is the best estimate of the activity of the epimerase (units (mg protein)$^{-1}$)?

(3) What is the smallest number of units of diaminopimelate dehydrogenase that should be added to a cuvette at pH 8.0 (3 ml total volume as before) to assay reliably 0.1 ml of a solution known to contain about 1 unit of diaminopimelate epimerase ml^{-1}?

The millimolar extinction coefficient of NADPH at 340 nm is 6.22 L $mmol^{-1}$ cm^{-1}, and may be assumed to be the same at pH 8.0 and at pH 10.5.

Problem 34

One more for fun. If this is data-handling then where are the data? They can be found. Think logically and mathematically. There is no catch, and it can be done.

Two mathematicians (Smith and Jones), who had been friends as undergraduates but had been out of touch for several years, were reunited at a meeting. The following conversation took place.

Jones: Since last we met I have married, and I have three children.

Smith: How old are they?

Jones: Work it out. The product of their three ages in years equals your own age.

Smith: I need more than that.

Jones: Right, the sum of their ages equals your room number in our hotel.

Smith: I still can't be sure.

Jones: What if I tell you that the eldest bites his nails?

Smith: Now I think I know.

Smith then gave the three ages correctly.

What were the ages of the children?

16 | Advice and hints

Below are Michelin star ratings for the difficulties of the problems:

* Easy, even though might be long
** Average, you should be able to cope
*** Hard, not obvious what to do, but not necessarily long

These ratings are based on my own opinions and on the experiences of students who have had to do battle with many of the problems. It's fairly certain that you will not find all the 'Easy' questions to be easy, but starting with them is still a good idea.

Comments after the ratings are intended as helpful hints

Problem 1   ** Note carefully the volumes of sulphuric acid and chloroform that are used.

Problem 2   * (some parts) ** (some parts) Read about logarithms.

Problem 3   **

Problem 4   * The x's cancel out.

Problem 5   *(*) Not quite so easy.

Problem 6   ** Just bash through – no great difficulties.

Problem 7   ** Read about enzymes and energy metabolism.

Problem 8   * It's tedious though.

Problem 9   *** The maths may be hard for some people. Read about logarithms.

Problem 10   *

Problem 11   * Divide % (w / w) by atomic weights.

Problem 12   *

Problem 13   **

Problem 14   **

Problem 15   *(*) Few get this right. Pick up clues in the introductory material.

Problem 16   *

Problem 17   *** No student can do this. It is essentially the glutamate synthase calculation (see Chapter 8: Enzymes) in reverse.

Problem 18   * Keep track of volumes: ml *of what* at all stages.

Problem 19   **(*)

Problem 20   **

Problem 21   *(*) Easy if you can do restriction mapping.

Problem 22   ** Long, but not hard if you follow the sequence of steps.

Problem 23   *(*)

Problem 24   *** Long, and a nasty one, but quite do-able.

Problem 25   ** Watch the units carefully.

Problem 26   **(*) A nice one.

Problem 27   *(*) Routine.

Problem 28   ** You need to know the formulae. Read Chapter 13 (Growth in continuous culture).

Problem 29   *(*)

Problem 30   **(*) Not too hard if you can do the mapping.

Problem 31   * Unfamiliar but easy (see Chapter 10: Energy metabolism).

Problem 32   *(*)

Problem 33   * The graph is really superfluous.

Problem 34   * or *** Depends on seeing what to do. What isn't Smith's age?

17 | Answers to problems

Calculation from graphs (Chapter 3)

The only real difficulty is in thinking how to make an equation that will allow you to solve the problem. Once the idea has come, the calculation is lengthy, but is just a lot of easy algebra.

The key is to say: let x minutes = number of minutes (after 180 min) at which strains A and B are present in equal numbers per ml. Then $x/45$ and $(x + 60)/30$ are the number of doublings of the two strains at $x + 180$ min.

Now we can write $1000 \times 2^{x/45} = 20 \times 2^{(x+60)/30}$ and can solve for x, provided that logarithms to base 2 are not too alarming.

Cancel the 20, so that $50 \times 2^{x/45} = 1 \times 2^{(x+60)/30}$

Now take logarithms to base 2:

$$\log_2 50 + x/45 = (x + 60)/30$$
$$\log_2 50 = (x + 60)/30 - x/45 = (3x + 180)/90 - 2x/90$$
$$= (x + 180)/90$$
$$90 \times \log_2 50 = x + 180$$
$$x = 90 \times \log_2 50 - 180$$
$$(\log_2 50 = \log_{10} 50/\log_{10} 2 = 1.6989/0.3010 = 5.644)$$
$$x = 90 \times 5.644 - 180 = 328 \text{ min}$$

Hence the time since inoculation when the two strains are equal is $328 + 180 = 508$ minutes.

The number of organisms of strain A per ml will then be $1000 \times 2^{328/45}$

$$= 1000 \times 2^{7.289}$$
$$= 1 \times 10^3 \times 1.56 \times 10^2 = 1.56 \times 10^5$$

The number of organisms of strain B should then be the same, so that the total organisms per ml at 508 minutes is 3.12×10^5

(Check: number of organisms of strain B per ml $= 20 \times 2^{388/30}$

$$= 20 \times 2^{12.93} = 20 \times 7.80 \times 10^3 = 1.56 \times 10^5.)$$

Now we can find the time (after 508 minutes) when strain B is in twofold excess of strain A. At this time (y min after 508 min) we shall have:

$$1 \times 2^{y/30} = 2 \times 2^{y/45}$$

Again taking logarithms to base 2 we get:

$$y/30 = 1 + y/45$$
$$1 = y/30 - y/45 = 3y/90 - 2y/90 = y/90$$

Hence $y = 90$ min

Strain B becomes twofold in excess of strain A at $508 + 90 = 598$ min
Strain A is just twofold in excess of strain B at $508 - 90 = 418$ min

The whole calculation can be done differently if you prefer to use the equation $n_x = n_0 e^{\mu x}$. Again x = minutes (after 180 min) at which the two strains are present in equal numbers.

Then we must first evaluate μ for each strain:

$$\mu_A = 0.693/45 = 0.0154 \text{ min}^{-1};$$
$$\mu_B = 0.693/30 = 0.0231 \text{ min}^{-1}$$

We can write $\ln n_A + \mu_A x = \ln n_B + \mu_B(x + 60)$ (n_A = initial number of strain A ml^{-1}; n_B = initial number of strain B ml^{-1}).

So $\ln 1000 - \ln 20 = \mu_B(x + 60) - \mu_A x$

$$6.9076 - 2.9957 = 0.0231(x + 60) - 0.0154x$$
$$3.9119 = 0.0231x + 1.386 - 0.0154x$$
$$3.9119 - 1.386 = 0.0077x$$
$$2.5259/0.0077 = x = 328 \text{ min}$$

Strain A count at 328 min $= n_{A328}$

$$\ln n_{A328} = (\ln 1000 + (\mu_A \times 328)) = (6.9076 + (0.0154 \times 328)) = 11.96$$
$$e^{11.96} = 1.56 \times 10^5 \, \text{ml}^{-1}$$

You can work out the rest if you want!
The graphical approach is easier, don't you think?

Problem 1

First, find out how much crotonic acid was present in the sulphuric acid.

A 1 M solution of crotonic acid in sulphuric acid has an absorbance of
1.56×10^4
Hence a 1 mM solution ($\equiv 1 \, \mu\text{mol ml}^{-1}$) has an absorbance of 1.56×10^1
Therefore an absorbance of 1.0 represents $1 / 15.6 \, \mu\text{mol ml}^{-1}$
And so an absorbance of 0.438 represents $0.438 / 15.6 \, \mu\text{mol ml}^{-1}$

Hence, in 10 ml of sulphuric acid there are $10 \times 0.438 / 15.6$ μmol crotonic acid = 0.281 μmol crotonic acid.

1 molecule of crotonic acid is equivalent to 1 repeating unit of poly
β-hydroxybutyrate
This repeat unit has a relative molecular mass of $(4 \times 12) + (6 \times 1) + (2 \times 16) = 86$ Da

Consequently, 0.281 μmol crotonic acid represents 0.281×86 μg poly β-hydroxybutyrate = 24.2 μg poly β-hydroxybutyrate.

The extract in chloroform therefore contained 24.2 μg poly
β-hydroxybutyrate in 1 ml
In 50 ml of chloroform there was 50×24.2 μg poly β-hydroxybutyrate =
1210 μg

This weight of poly β-hydroxybutyrate came from 5 ml of bacterial
suspension

20 ml of this suspension had a dry wt of 24 mg
5 ml therefore had a dry wt of 6 mg

Now we can say that 6 mg dry wt of bacteria contained 1.21 mg poly
β-hydroxybutyrate

$\%(\mathrm{w/w})$ **poly β-hydroxybutyrate** $= 1.21 \times 100/6 = 20.2\%$

(This result is typical for *Bacillus megaterium* grown on glucose as sole source of carbon. Not many species make as much poly β-hydroxybutyrate as can *Azotobacter* spp.)

Problem 2

(1) x

(2) $2x / y^2$

(3) x^{12}

(4) $27x^9$

(5) x^2

(6) $x^{-1.25}$

(7) $^2\sqrt{x}$

(8) $(x^2 / y^2) + 1$

(9) 4

(10) 5

(11) 7

(12) 4

(13) 0.954

(14) 1.079

(15) −0.125

(16) 0.3597

(17) 0.8405

(18) 0.778

(19) $x = c^y$

(20) $x = c^{y/2}$

(21) $x = {}^y\sqrt{c}$ (because $x^y = c$)

(22) $x = c^{2y}$ (because $2y = \log_c x$)

(23) $x = \log_c y$ and $x = \ln y / \ln c$ (because $\ln y = x \ln c$); hence $\log_c y = \ln y / \ln c$

(24) 5.1699

(25) −0.4112

(26) 13.235

(27) 0.3515

(28) 1.8369

(29) 1.1892

(30) 16

(31) 19.8130

(32) 1.4186

(33) £162.89; 164 organisms (or 165; 164.87 is the precise answer). The two answers differ because the interest is given in 10 discrete steps, while the culture, though growing at the same percentage rate as the money, increases continuously.

(34) 1683 dpm; decay constant $= 0.0495$ day^{-1}

(35) Doubling time $= 1$ h

$\mu = 0.693$ h^{-1}

The exponential phase ends at approx. 630 min; the exponential phase begins sometime before 290 min of incubation

(36) (a) 8.5 min; (b) 2.7 min; (c) 0.023%

(37) -0.602 (4 M H$^+$)

(38) 13.48

(39) 0.824

(40) 31.

Problem 3

(1)

Glucose-grown organisms

Suspension

1 ml suspension (undiluted) hydrolysed 1.2 μmol ONPG in 30 min

So, **1 ml suspension (undiluted) hydrolyses 0.04 μmol ONPG min^{-1}**

0.5 ml suspension (diluted 1 : 50) contained 46 μg protein

∴ 0.5 ml suspension (undiluted) contained 2300 μg protein

∴ **1 ml suspension (undiluted) contains 4.6 mg protein**

0.04 μmol ONPG hydrolysed by 1 ml suspension ($\equiv$ 4.6 mg protein) min^{-1}

∴ **With suspension, 0.0087 μmol ONPG hydrolysed min^{-1} (mg protein)$^{-1}$**

Extract

1 ml extract (undiluted) hydrolysed 0.8 μmol ONPG in 10 min

So, **1 ml extract (undiluted) hydrolyses 0.08 μmol ONPG min^{-1}**

0.5 ml extract (diluted 1 : 100) contained 20 μg protein

∴ 0.5 ml extract (undiluted) contained 2000 μg protein

∴ **1 ml extract (undiluted) contains 4 mg protein**

0.08 μmol ONPG hydrolysed by 1 ml extract ($\equiv$ 4.0 mg protein) min^{-1}

∴ **With extract, 0.02 μmol ONPG hydrolysed min^{-1} (mg protein)$^{-1}$**

Lactose-grown organisms   Similar reasoning to the above gives:

Suspension   **0.02 μmol ONPG hydrolysed min^{-1} (mg protein)$^{-1}$**

Extract   **2.5 μmol ONPG hydrolysed min^{-1} (mg protein)$^{-1}$**

(2) In the suspension of glucose-grown organisms the specific activity of β-galactosidase (μmol ONPG hydrolysed min^{-1} (mg protein)$^{-1}$) is low (0.0087) while in the extract of these organisms the specific activity is somewhat greater (0.02). This suggests that the substrate does not readily enter the intact bacteria, so that the true specific activity is not shown when the organisms are undamaged.

    In the suspension of lactose-grown organisms the specific activity of β-galactosidase is again low (0.02) though higher than in the suspension of glucose-grown organisms (0.0087). The specific activity of β-galactosidase in the extract of lactose-grown organisms (2.5) is 100 times more than in the extract of glucose-grown bacteria. This indicates that the enzyme is induced by lactose, which, in order to be metabolised, must first be hydrolysed to glucose and galactose by β-galactosidase. Impermeability to ONPG prevents the activity of β-galactosidase being fully revealed when the organisms in suspension are unbroken.

Problem 4

(1) From 1 L of culture there are $2 \times 10^9 \times 10^3$ organisms, which weigh 2 g

     Hence 1 organism weighs 1×10^{-12} g

     1×10^x g organisms contain 6×10^{23} molecules

     1 g organism contains $6 \times 10^{(23-x)}$ molecules

     1×10^{-12} g organisms (one organism) contains $6 \times 10^{(11-x)}$ molecules

     Each molecule contains $1 \times 10^{(x-1)}$ atoms

     So 1 organism is made up of approx. $6 \times 10^{(11-x)} \times 1 \times 10^{(x-1)}$ atoms $= 6 \times 10^{10}$ atoms

2 (a) Internal volume $= 2 \times \pi \times 0.25^2 \ \mu m^3 = 0.125 \ \pi \ \mu m^3 = 0.4 \ \mu m^3$

$1 \mu m = 1 \times 10^{-6} \ m$

Hence internal volume $= 0.4 \times 10^{-18} \ m^3 = 4 \times 10^{-19} \ m^3$

(b) A 1 M solution of H^+ ions in water contains $6 \times 10^{23} \ H^+$ ions L^{-1}

At pH 7 ($H^+ = 1 \times 10^{-7} \ M$) there are $6 \times 10^{16} \ H^+$ ions L^{-1}

and $1 \ L = 1 \times 10^{-3} \ m^3$

So, at pH 7 there are $6 \times 10^{19} \ H^+$ ions per m^3

Thus in $1 \times 10^{-19} \ m^3$ there are 6 H^+ ions at pH 7

In one organism there are $4 \times 6 \ H^+$ ions $= 24 \ H^+$ ions

Problem 5

2.75×10^3 dpm $\equiv 1 \ \mu g \ ^3H$ thymidine

$\therefore 2.0 \times 10^3$ dpm taken up $\equiv 2.0/2.75 \ \mu g$ thymidine

Hence, weight of staphylococci in film $= (2.0 \times 100) / (2.75 \times 0.3) \ \mu g = 242 \ \mu g$

This is the same as $242 \times 10^{-6} \ g$

$\therefore$ Number of organisms in film $= (242 \times 10^{-6}) / (5 \times 10^{-13}) = 48.4 \times 10^7 =$

4.84×10^8

$7 \ mm = 7000 \ \mu m$ and $25 \ mm = 25\,000 \ \mu m$, so area of slide $= 7 \times 10^3 \times 25 \times$

$10^3 = 175 \times 10^6 \ \mu m^2$

Number of spheres (radius $= 0.5 \ \mu m$) that can pack this area is $(1.75 \times 10^8) /$

$(0.5^2 \times 2 \times \sqrt{3})$

$= 2.02 \times 10^8$

$\therefore$ **Number of layers formed by 4.84×10^8 organisms $= (4.84 \times 10^8) /$**

$(2.02 \times 10^8) = 2.4$

One of these layers is bottom, attached to glass, so that **% attached $= 1 \times 100 /$**

$2.4 = 41 \ \%$

Reasons for overestimate of fraction (%) attached (in probable decreasing order of importance):

(i) The attached organisms are very unlikely to achieve an even, close-packed distribution over the slide.

(ii) If the organisms continue to synthesise (unlabelled) thymidine from its precursors when ^3H-thymidine is in the medium then the incorporated dpm will be an underestimate of the total thymidine (labelled and unlabelled) in the biofilm, and so an underestimate of the total number of bacteria in the film.

(iii) Incorporated ^{3}H-thymidylic acid will be in DNA in the middle of each organism, so that there may be some attenuation of the true count (owing to self-absorption) which means that the amount of ^{3}H-thymidylic acid inside the bacteria (and so the organisms themselves) will once again be underestimated.

Problem 6

(i) Medium A contains 20 mM fructose and 2.0 g $(NH_4)_2SO_4$ L^{-1}

1 mM fructose $\equiv$ 180 mg fructose L^{-1}

20 mM fructose $\equiv$ 3.6 g fructose L^{-1}

1 g fructose can yield 0.7 g bacteria

$\therefore$ 3.6 g fructose can yield 2.52 g bacteria

The fructose in 1 L of medium A can yield 2.52 g bacteria

The mol. wt of ammonium sulphate is $(18 \times 2) + 32 + (4 \times 16)$

$= 132$

132 g ammonium sulphate contain 28 g nitrogen

1 g ammonium sulphate contains 28 / 132 g nitrogen

2.0 g ammonium sulphate contain $28 \times 2 / 132$ g nitrogen $= 0.42$ g nitrogen

15 g nitrogen can yield 100 g bacteria

1 g nitrogen can yield 100 / 15 g bacteria $= 6.67$ g bacteria

0.42 g nitrogen can yield 6.67×0.42 g bacteria $= 2.8$ g bacteria

The ammonium sulphate in 1 L of medium A can yield 2.8 g bacteria

Thus, the medium is carbon-limited for growth.

(2) Tris mol. wt $= 16 + 12 + (31 \times 3) = 28 + 93 = 121$

121 g L^{-1} = 1 M; 121 / 4 g L^{-1} = 250 mM

121/4 ÷ 200 g in 5 ml = 250 mM

0.2 ml of this stock solution made up to 1 ml will be 50 mM Tris

$MgCl_2$ mol. wt $= 24 + 71 = 95$

95 mg L^{-1} = 1 mM : 95×50 mg L^{-1} = 50 mM

$95 \times 50 ÷ 200$ mg (= 23.8 mg) in 5 ml = 50 mM

0.1 ml of this stock solution made up to 1 ml will be 5 mM $MgCl_2$

EDTA mol. wt = $[14 + 14 + (14 + 28 + 16 + 1)_2]_2 = [28 + (59 \times 2)]_2$
$= (28 + 118) \times 2 = 146 \times 2 = 292$

$292 \, \text{mg L}^{-1} = 1 \, \text{mM} \equiv 29.2 \, \text{mg in } 100 \, \text{ml}$

Take 0.5 ml of this solution and make up to 5 ml (= 0.1 mM)

0.1 ml of this stock solution made up to 1 ml will be 0.01 mM EDTA

NADP mol. wt = $92 + (12 \times 21) + 26 + (7 \times 14) + (17 \times 16) + 93$
$= 92 + 252 + 26 + 98 + 272 + 93 = 833$

$833 \, \text{mg L}^{-1} = 1 \, \text{mM}; \, 4 \times 833 \, \text{mg L}^{-1} = 4 \, \text{mM}$

$4 \times 833 \div 200 \, \text{mg} \, (= 16.7 \, \text{mg}) \, \text{in } 5 \, \text{ml} = 4 \, \text{mM}$

0.1 ml of this stock solution made up to 1 ml will be 0.4 mM NADP

Linkage enzyme solution is 300 units per ml

Take 0.5 ml and make up to 5 ml with appropriate buffer = 30 units ml^{-1}

0.1 ml of this stock solution made up to 1 ml will give 3 units

Fructose-1,6-bisphosphate mol. wt = $72 + 11 + (12 \times 16) + 62 + 69 + (8 \times 18) = 83 + 192 + 62 + 69 + 144 = 550$

$550 \, \text{mg L}^{-1} = 1 \, \text{mM}; \, 550 \times 20 \, \text{mg L}^{-1} = 20 \, \text{mM}$

$550 \times 20 \div 200 \, \text{mg} \, (= 55 \, \text{mg}) \, \text{in } 5 \, \text{ml will be } 20 \, \text{mM}$

0.1 ml of this stock solution made up to 1 ml will be 2 mM fructose-1,6-bisphosphate

(3) An extinction of $6.22 \times 10^3 \equiv 1 \, \text{M NADPH} \equiv 1 \, \text{mmol ml}^{-1}$

$\therefore 6.22 \equiv 1 \, \mu\text{mol ml}^{-1}$ and extinction of $1 \equiv 1/6.22 \, \mu\text{mol ml}^{-1}$

so that extinction $x \equiv x/6.22 \, \mu\text{mol ml}^{-1}$

and $0.209 \equiv 0.209/6.22 \, \mu\text{mol ml}^{-1} = 209/6.22 \, \text{nmol ml}^{-1} = 33.6 \, \text{nmol ml}^{-1}$

This amount is formed in 1 min by 0.01 ml extract

1 ml extract forms 3360 nmol / min

25 ml extract makes 84 000 nmol / min

Specific activity = $3360 / 1.9 = 1770 \, \text{nmol min}^{-1} \, (\text{mg protein})^{-1}$

The activities of the two supernates can easily be found by proportion:

e.g. $\Delta_{340 \, \text{nm}}$ of $0.121 \equiv 33.6 \times 121/209 = 19.5 \, \text{nmol ml}^{-1}$

Treatment of extract	Vol. in assay (ml)	Total vol. (ml)	Δ_{340nm} optical density min^{-1}	Activity (nmol min^{-1} ml^{-1})	Total activity (nmol min^{-1})	Specific activity (nmol min^{-1} (mg protein)$^{-1}$)
Crude extract	0.01	25	0.209	**3360**	**84 000**	1770
Acid supernate	0.01	27.5	0.121	**1950**	**53 600**	1770
$(NH_4)_2SO_4$ supernate	0.02	30.5	0.151	**1215**	**37 100**	2530

Problem 7

(1) 1 M MV⁻ has an extinction of 7.4×10^3

$\therefore$ extinction of $1 = 1/(7.4 \times 10^3)$ mol L^{-1}

$\equiv 1/(7.4 \times 10^3)$ mmol ml^{-1}

$1/7.4\,\mu$mol ml^{-1}

A change of extinction of 0.18 $min^{-1} \equiv 0.18 / 7.4$ µmol ml^{-1} min^{-1}

$\therefore$ in 3 ml (cuvette contents) 0.54 / 7.4 µmol formed min^{-1} by 0.1 ml extract

Hence, 1 ml extract forms 5.4 / 7.4 µmol min^{-1} = 730 nmol min^{-1}

The extract contained 14.3 mg protein ml^{-1}

So, **specific activity of the enzyme is 730 / 14.3 = 51 nmol min^{-1} (mg protein)$^{-1}$**

$(\equiv \textbf{0.051 unit (mg protein)}^{-1})$

(2) 1 mg of purified enzyme contains 0.97 µg nickel

1 g will contain 0.97 mg Ni

1 kg will contain 0.97 g Ni

60 kg (one mole of this enzyme) will contain 60×0.97 g Ni = 58 g

This is very nearly = 1 g atom Ni

One molecule of the enzyme contains 1 atom of nickel

(3) Specific activity of pure enzyme is 100 × that of the extract

and so is 5.1 µmol (mg protein)$^{-1}$ = 5.1 units / (mg protein)$^{-1}$

Assume that there is one active site (one atom of nickel per molecule of enzyme)

∴ **Turnover number** $= 5.1 \times 60\,000 \times 10^{-3} / 60 \times 1 = \textbf{5.1 molecules s}^{-1}$

(4) We have two half-reactions:

$$CO + H_2O \rightarrow CO_2 + 2H^+ + 2e^-$$

and

$$2MV + 2H^+ + 2e^- \rightarrow 2MV^- + 2H^+$$

Overall $CO + H_2O + 2MV \rightarrow CO_2 + 2H^+ + 2MV^-$

$\Delta E_0'$ for this is $+ 0.09$ V, and two electrons are transferred

∴ $\Delta G_0' = -2 \times 0.09 \times 23\,061$ cal $= \textbf{4159 cal} = \textbf{4.2 kcal (mol CO)}^{-1}$

Bear in mind the natural final electron-acceptor is likely to have a more positive E_0' value than does MV, so that the yield of energy may be greater, per mole of CO oxidised, than is shown by the above calculation.

Problem 8

(1) $Na_2SO_4.10H_2O$ mol. wt $= (22.9 \times 2) + 32 + (16 \times 4) + (10 \times 18) = 321.8$

Therefore 321.8 g $Na_2SO_4.10H_2O$ contains 32 g S

And 321.8 / 32 g $Na_2SO_4.10H_2O$ contains 1 g S

Hence 10.06 g $Na_2SO_4.10H_2O$ in 1 L will have 1 g S $L^{-1} = 1$ mg S ml$^{-1} =$ 1000 μg S ml^{-1}

1.006 g $Na_2SO_4.10H_2O$ in 1 L will give 100 μg S ml^{-1}. Call this solution A. For standard curve prepare:

0.2 ml A + 0.8 ml water (20 μg S)

0.4 ml A + 0.6 ml water (40 μg S)

0.6 ml A + 0.4 ml water (60 μg S)

0.8 ml A + 0.2 ml water (80 μg S)

1.0 ml A   (100 μg S)

$Na_2S_2O_3.5H_2O$ mol. wt $= 247.8$, so that 247.8 g contains 64 g S

And 247.8/64 g contains 1 g S

3.872 g L^{-1} gives 1 g S L^{-1}

387 mg in 1 L will give 100 μg S ml^{-1}. Call this solution B.

Use solution B as above (in place of A) to make a standard curve.

(2) Plotting the data for the standard curves (OD against μg S ml^{-1}) gives two straight-line graphs, from which one can read off values for μg S ml^{-1} from the OD readings of the test samples:

SO_4^{2-}-S (μg S ml^{-1})	$S_2O_3^{2-}$-S (μg S ml^{-1})
85	90
100	85
95	85
90	79

SO_4^{2-}-S in soil

1 g soil in 25 ml, 1 ml of this made up to 5 ml.

$\therefore$ 1 ml assayed represents 1/125 g soil

$\therefore$ 93 μg S (average of 4 estimates) in 1/125 g soil

93 $\times$ 125 μg S in 1 g soil = 11 625 μg

11.6 mg S as SO_4^{2-} per gram of soil

$S_2O_3^{2-}$-S in soil

5 g soil in 20 ml, 1 ml of this made up to 5 ml.

$\therefore$ 1 ml assayed represents 5/100 g soil

$\therefore$ 85 μg S (average of 4 estimates) in 1/20 g soil

85 $\times$ 20 μg S in 1 g soil = 1700 μg

1.7 mg S as $S_2O_3^{2-}$ per gram of soil

Problem 9

When V_w = 10 ml then:

$$C_w = 0.5 - 0.5/e^{(10/50)} = 0.5 - 0.5/1.221 = 0.5 - 0.409$$
$$= \mathbf{0.091} \text{ M}$$

Similarly,

When V_w = 20 ml then C_w = 0.165 M
When V_w = 30 ml then C_w = 0.226 M
When V_w = 50 ml then C_w = 0.316 M
When V_w = 75 ml then C_w = 0.388 M
When V_w = 100 ml then C_w = 0.432 M
When V_w = 150 ml then C_w = 0.475 M

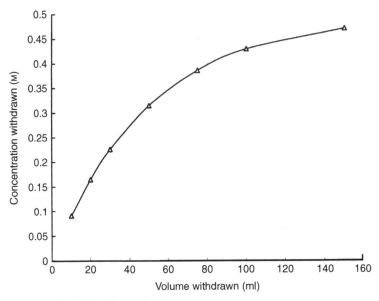

Fig. 17.1 The gradient is convex.

Now a graph can be drawn (see Fig. 17.1).
When the concentration withdrawn is 0.25 M then

$$V_w = 50 \times \ln [0.5/(0.5 - 0.25)] = 50 \times \ln (0.5/0.25) = 50 \times \ln 2$$
$$= 50 \times 0.693 = 34.7 \, ml$$

The value of V_w read from the graph should be approximately the same.

$$C_w = C_R - C_D \, e^{-(V_w/V_m)}$$

Rearranging, $C_D/e^{(V_w/V_m)} = C_R - C_w$
Multiply both sides of this equation by $e^{(V_w/V_m)}$

$$C_D = (C_R - C_w) \times e^{(V_w/V_m)}$$

Divide both sides by $(C_R - C_w)$

$$C_D/(C_R - C_W) = e^{(V_w/V_m)}$$

Hence $\ln [C_D / (C_R - C_w)] = V_w / V_m$
And so

$$V_w = V_m \times \ln [C_D/(C_R - C_w)]$$

The deceptive part of this last calculation is that you need to realise that fraction 3 (and not fraction 1 or 2) will contain the first 5 ml of the NaCl gradient. This is a consequence of the void volume of the column, which is 10 ml (equal to two fractions of 5 ml).

Therefore, fractions 3 to 6 represent the first 20 ml of the gradient, which did not elute the enzyme. When V_w is 20 ml, then C_w will be:

$$0.4 - 0.3/e^{(20/30)} = 0.4 - 0.3/1.948 = 0.4 - 0.154 = 0.246 \text{ M}$$

By a similar calculation when V_w is 25 ml (fraction 7) C_w will be 0.270 M Hence, 0.246 M NaCl was the highest concentration that did not elute the enzyme.

And 0.270 M NaCl was the lowest concentration that did elute the enzyme.

Problem 10

It's all too easy to be worth showing the calculations

By solving the equation $\mu_{max} = 1.00$
Then we can find $K_s = 1.1$
From the graph (not shown) $\mu_{max} = 1.0$

$$K_s = 1.10$$

Problem 11

(1) First divide the % (w / w) of each element by its atomic weight to find the number of atoms in the empirical formula:
Carbon 48% / 12 = 4
Oxygen 32% / 16 = 2
Nitrogen 14% / 14 = 1
Hydrogen 6% / 1 = 6
The empirical formula is $C_4H_6O_2N$

(2) We have glucose + ammonia + oxygen being converted to $C_4H_6O_2N$ + $CO_2 + H_2O$

$$C_6H_{12}O_6 + NH_3 + xO_2 \rightarrow C_4H_6O_2N + 2CO_2 + 4\tfrac{1}{2}H_2O$$
$$= 2C_6H_{12}O_6 + 2NH_3 + xO_2 \rightarrow 2C_4H_6O_2N + 4CO_2 + 9H_2O$$

On the left we have $12O + x\,O_2$ and on the right $21O$
Hence, $x = 4\frac{1}{2}$
To remove fractions we can now write:

$$4C_6H_{12}O_6 + 4NH_3 + 9O_2 \rightarrow 4C_4H_6O_2N + 8CO_2 + 18H_2O$$

(3) $4C_4H_6O_2N$ has a weight of $4 \times (48 + 6 + 32 + 14)$ Da $= 400$ Da
So, 400 g biomass per 9 moles O_2 consumed
Hence $Y_{O_2} = 400 / 9 = 44.4$ g (mole $O_2)^{-1}$

(4) 18 atoms of oxygen are consumed in producing 400 g biomass
So 400 g biomass per 36 moles ATP
Hence $Y_{ATP} = 400 / 36 = 11.1$ g mole^{-1}

Problem 12

1 M ATP has OD 1.54×10^4 at 259 nm and pH 7
1 mM ATP has OD $15.4 \equiv 1$ μmole ml^{-1}
$\therefore$ OD $1.0 = 1 / 15.4$ μmole ml^{-1}

So, OD $0.45 = 0.45 / 15.4$ μmole ml$^{-1} = 0.0292$ μmole ml^{-1}
Solution B contains 0.0292 μmole ml^{-1}
$\therefore$ Solution A contains 2.92 μmole ml^{-1}
And 10 ml solution A contains 29.2 μmoles

$$29.2\ \mu\text{moles} \equiv 507 \times 29.2\ \mu\text{g ATP} = 14\,800\ \mu\text{g} = 14.8\ \text{mg}$$
$$\therefore \%\ \textbf{purity} = (14.8/20) \times 100 = 74\%(\text{w}/\text{w})$$

To make 50 ml of 1 mM ATP we need 50 μmoles
50 / 2.92 ml A $= 17.1$ ml gives 50 μmoles
Take 17.1 ml A plus 32.9 ml water

Solution A contains 2.92 μmoles ATP ml^{-1} = 8.76 μmoles P

$$\equiv 31 \times 8.76\ \mu\text{g P} = 272\ \mu\text{g P ml}^{-1}$$

0.1 ml solution A contains 27.2 μg P
Dilute solution A 1 ml + 9 ml water (1 : 10) to get solution C containing 27.2 μg P ml^{-1}
There will be 2.72 μg P in 0.1 ml solution C

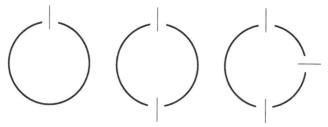

Fig. 17.2 Breakage of a circular plasmid at one site will produce one (linear) fragment. Breakage at two sites will yield two fragments; three sites will yield three fragments, and so on.

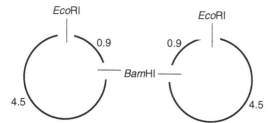

Fig. 17.3 Relation of *Eco*RI and *Bam*HI sites on pLX100.

Problem 13

(1) The number of fragments arising from a circular plasmid will equal the number of sites at which restriction endonucleases have attacked (Fig. 17.2). Thus, there must be one *Eco*RI site, one *Bam*HI site and two *Pst*I sites on plasmid pLX100.

The *Eco*RI + *Bam*HI digest shows that the sites of action of these two enzymes are 0.9 kbp apart (Fig. 17.3). Note that either representation is equally good, and so do not become fixed on just one of these at this stage.

The *Pst*I digest contains two fragments (3.3 kbp and 2.1 kbp), so that this enzyme must have two sites of action. These two sites have now to be located in their correct relationships to the *Eco*RI and *Bam*HI sites.

When *Pst*I is used with *Eco*RI the 3.3 kbp fragment disappears, and is replaced by 2.3 kbp and 1.0 kbp fragments, while the 2.1 kbp fragment remains. Since *Eco*RI has only one site of action, we can now deduce that the *Eco*RI site is 1.0 kbp away from one of the *Pst*I sites, and there are two possible positions for the *Bam*HI site (Fig. 17.4).

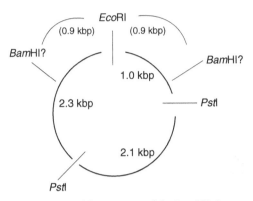

Fig. 17.4 Possible positions of the *Bam*HI site.

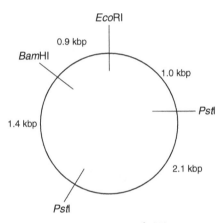

Fig. 17.5 Restriction map of pLX100.

If the *Bam*HI site were to the right of the *Eco*RI (in Fig. 17.4) then the *Bam*HI + *Pst*I digest would have contained fragments of sizes 0.1, 3.2 and 2.1 kbp, which is not what was found. A site to the left of *Eco*RI yields fragments of sizes 1.9, 1.4 and 2.1 kbp, which is what was found. The restriction map must therefore be as shown in Fig. 17.5.

(2) 2 µl of a suspension containing 100 µg ml^{-1} will contain 0.2 µg pLX100. Next there are 0.2 µg in 1 ml, so that 2 µl of this suspension will contain 0.0004 µg ≡ 0.4 ng

0.1 ml of bacterial suspension taken from a total volume of 1.102 ml gave 250 colonies

Hence, total transformants = 250 × 11.02 = 2755

So, 0.4 ng plasmid transformed 2755 organisms ≡ 6888 transformed per ng plasmid **and 6.89 × 10^6 transformed per μg plasmid**

(3) pLX100 contains 5.4 kbp ≡ 5.4 × 10^3 × 2 × 330 Da per plasmid = 3.56 × 10^6 Da

6.02 × 10^{23} plasmids will therefore weigh 3.56 × 10^6 g

Therefore 1 g plasmid will contain (6.02 × 10^{23})/(3.56 × 10^6) particles = 1.69 × 10^{17} plasmids

And 1 μg plasmid will contain 1.69 × 10^{11} plasmids

So we have 6.89 × 10^6 transformants from 1.69 × 10^{11} plasmids

$$\equiv (6.89 \times 10^6 \times 100)/(1.69 \times 10^{11})\% \textbf{ of plasmids recovered as}$$

transformants = 4.08 × 10^{-3}%

Problem 14

The mol. wt of one unit of polysaccharide polymer is 162 + 146 + 146 + 176 = 630 Da

$$400 \text{ kBq} = 2.4 \times 10^7 \text{ dpm}$$

2.4 × 10^7 dpm represents 100 mg diaminopimelate (= 100 / 190 mmole)

Hence 1 × 10^7 dpm represents 0.219 mmole diaminopimelate (= 219 μmole)

And 1 × 10^3 dpm represents 0.0219 μmole diaminopimelate

Therefore 1 ml of polysaccharide solution contains 0.0438 μmole diaminopimelate

And 1 ml of polysaccharide solution also contains 40 μmole rhamnose

For 1 molecule of diaminopimelate there are 40 / 0.0438 = 914 molecules of rhamnose

914 molecules of rhamnose mean 457 polymer units of polysaccharide

Hence the apparent mol. wt is 457 × 630 = 288 000 Da

Problem 15

First we need to find the numbers of survivors per ml of the suspension at 80 °C after various times of exposure to this temperature.

After 1 min at 80 °C

240 colonies came from 0.1 ml of the buffer suspension at 30 °C
So 2.4×10^3 colonies came from 1 ml of this suspension
and 2.4×10^4 colonies came from 10 ml of this suspension.
This represents the survivors from 0.1 ml of the suspension at 80 °C
Hence there were 2.4×10^5 survivors per ml of the suspension at 80 °C

$\log_{10}$ (survivors per ml at 80 °C) = 5.38

After 2 min at 80 °C

407 colonies came from 0.5 ml of the buffer suspension at 30 °C
So 814×10^2 colonies came from 1 ml of this suspension
and 8.14×10^3 colonies came from 10 ml of this suspension.
This represents the survivors from 0.1 ml of the suspension at 80 °C
Hence there were 8.14×10^4 survivors per ml of the suspension at 80 °C

$\log_{10}$ (survivors per ml at 80 °C) = 4.91

Results for the other times can be worked out similarly:

After 3 min $\log_{10}$ (survivors per ml at 80 °C) = 4.39
After 4 min $\log_{10}$ (survivors per ml at 80 °C) = 3.81
After 6 min $\log_{10}$ (survivors per ml at 80 °C) = 3.48
After 8 min $\log_{10}$ (survivors per ml at 80 °C) = 3.60
After 12 min $\log_{10}$ (survivors per ml at 80 °C) = 3.79
After 18 min $\log_{10}$ (survivors per ml at 80 °C) = 3.78
After 30 min $\log_{10}$ (survivors per ml at 80 °C) = 3.76

(1) The required graph now can be plotted (Fig. 17.6).

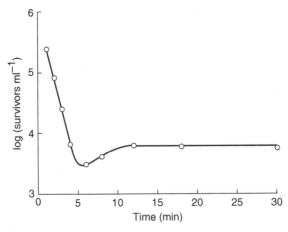

Fig. 17.6

(2) The key to this question is the word 'spore'. The culture was isolated from pasteurised soil (which procedure strongly selects for spore-formers); in the defined medium the organisms have limited carbon (which promotes sporulation); the culture was sampled late in the exponential phase of growth (which is when spores begin to be formed by the majority of the organisms). In spite of these clues about 90% of the students who attempt the problem have failed to think of spores at all, and this makes it difficult to achieve a simple or plausible interpretation of the results.

Interpretation   The organisms are spore-formers. Sporulation is beginning in the culture in the defined medium when the exponential phase approaches its end because of starvation of carbon.

Exposure to 80 °C rapidly kills the vegetative bacteria that have not yet formed spores. The decimal reduction time (the time needed to decrease the number of viable organisms by a factor of 10) is approximately 2 min at 80 °C. The spores survive at 80 °C for at least 30 min.

The rise in the number of survivors (after the minimum at 6 min) occurs because these spores require a short exposure to a high temperature (about 10 min at 80 °C) in order to be able to germinate and hence produce colonies. Heat activation is often seen as a property of bacterial spores.

(3) To estimate the number of organisms in the culture at time 0 one may most logically calculate from the number of survivors after 1 min at 80 °C. We do not know for certain that there is any killing during the first minute. Almost equally reasonably one may assume that killing starts at once, and so extrapolate the straight line (logarithmic death) upwards until it intercepts the *y* axis, and then use this intercept value in the calculation. Either method would give an acceptable answer to this question.

Let's just do the (very slightly) more complicated calculation from the intercept.

The intercept is at approximately $\log_{10}$ (viable organisms per ml at 80 °C) = 5.9

Hence there are 7.9×10^5 organisms per ml at 80 °C

And so 7.9×10^7 organisms in 100 ml at 80 °C

These organisms came from 0.1 ml of the culture at 30 °C

The culture contained 7.9 × 10^8 organisms per ml at time 0.

(4) After 30 min at 80 °C there were 5.8 ×10^3 colonies from 1 ml of the suspension at 80 °C

At time 0 there were 7.9 × 10^5 colonies from 1 ml of the suspension at 80 °C

Hence % surviving at 80°C for 30 min $= (10^2 \times 5.8 \times 10^3)/$
$$(7.9 \times 10^5) = 0.73\%$$

(If we wanted to be exquisitely fussy we could say that the count at time 0 might include only vegetative organisms because the spores would perhaps not have germinated without exposure to 80 °C. The total count per ml of culture should then be 7.9 × 10^8 + 5.8 × 10^6 which is 7.9 × 10^8 + 0.058 × 10^8 = <8.0 × 10^8. Thus, adding the spore count makes very little change in the total count at time 0.)

Problem 16

(1) Average colonies per plate for soil A = 8.4

These were present in 0.1 ml of the suspension plated

So, 8.4 × 10^1 ml^{-1} of suspension plated

And hence 8.4 × 10^4 ml^{-1} of suspension in Ringer's solution

10 ml of Ringer's solution used to suspend organisms from 1 g of soil A

Therefore 8.4 × 10^5 thiobacilli per g of soil A

Average colonies per plate for soil B = 6.0

By similar reasoning there are 6 × 10^5 thiobacilli per g of soil B

(2) **For soil A median = 9; mean = 8.4**

$$\text{Variance}_A = [(10^2 + 9^2 + 9^2 + 8^2 + 6^2)/5] - 8.4^2$$
$$= [(100 + 81 + 81 + 64 + 36)/5] - 70.56$$
$$= (362/5) - 70.56 = 72.4 - 70.56 = 1.84$$

Sample variance = 1.84 × 5 / 4 = 2.30

Sample standard deviation (s_A) = $\sqrt{2.30}$ = 1.52

For soil B median = 8; mean = 6.0

$$\text{Variance}_B = [(8^2 + 8^2 + 8^2 + 3^2 + 3^2)/5] - 6^2$$
$$= [(64 + 64 + 64 + 9 + 9)/5] - 36$$
$$= (210/5) - 36 = 42 - 36 = 6$$

Sample variance = $6 \times 5 / 4 = 7.5$
Sample standard deviation (s_B) = $\sqrt{7.5} = 2.74$

t-*test*

$$t = |\bar{x}_1 - \bar{x}_2|/[s_p\sqrt{(1/n_1 + 1/n_2)}]$$

where s_p is the square root of $[(\, n_1 - 1)s_1{}^2 + (n_2 - 1)s_2{}^2] / (n_1 + n_2 - 2)$
For these two soils $|\bar{x}_1 - \bar{x}_2| = 8.4 - 6.0 = 2.4$

$$s_p = \sqrt{\{[(5 - 1) \times 2.30 + (5 - 1) \times 7.5] / (5 + 5 - 2)\}}$$
$$= \sqrt{\{[(4 \times 2.30) + (4 \times 7.5)] / 8\}}$$
$$= \sqrt{[(9.2 + 30)/8]} = \sqrt{(39.2/8)} = \sqrt{4.9} = 2.21$$
$$\sqrt{(1/5 + 1/5)} = \sqrt{0.4} = 0.632$$

Now we can write t = 2.4 / (2.21 × 0.632) = 2.4 / 1.40 = 1.71

$$t = 1.71$$

From statistical tables we see that when n_1 and n_2 both = 5 then t must be $\not>2.306$ before mean A can be greater than mean B with 95% probability.
The counts of the soil A are not significantly higher than those of soil B.

Mann–Whitney test   Rank the 10 counts in descending order and show for each count whether it was for soil A or soil B, and then number from 10 to 1 downwards:

10	9	9	8	8	8	8	6	3	3
A	A	A	A	B	B	B	A	B	B
10	8.5	8.5	5.5	5.5	5.5	5.5	3	1.5	1.5

Add the A numbers: $10 + 8.5 + 8.5 + 5.5 + 3 = 35.5 = W_1$
Add the B numbers: $5.5 + 5.5 + 5.5 + 1.5 + 1.5 = 19.5 = W_2$

$$U_1 = W_1 - n_1(n_1 + 1)/2 = 35.5 - (5 \times 6)/2 = 35.5 - 15 = 20.5$$
$$U_2 = W_2 - n_2(n_2 + 1)/2 = 19.5 - (5 \times 6)/2 = 19.5 - 15 = 4.5$$
$$U = \text{the smaller of } U_1 \text{ and } U_2 = 4.5$$

Since U is >2 (see Chapter 6) the counts of soil A are not significantly higher than the counts of soil B.

Problem 17

To find this final specific activity, first work out the specific activity of isocitrate lyase during growth on acetate.
  Here we go:

An absorbance change of 0.863 in 5 min is caused by 75 µl of extract
  (diluted 50-fold)
As this change had progressed linearly, this is equivalent to 0.1726 in 1 min
An absorbance of 1.7×10^4 is given by a 1 M solution of glyoxylate
  phenylhydrazone
Therefore, an absorbance of 1.7×10 will be given by a 1 mM solution
And so an absorbance of 1.0 will represent a 1 / 17 mM solution

$$(\equiv 1/17 \text{ mmol } L^{-1} \equiv 1/17 \text{ µmol ml}^{-1})$$

Hence, an absorbance of 0.1726 represents 0.1726 / 17 µmol ml^{-1}
There was 3 ml in the cuvette, and so $3 \times 0.1726 / 17$ µmol glyoxylate was
  produced in 1 min by 75 µl of extract (diluted 50-fold), i.e. 0.030 46 µmol
  glyoxylate min^{-1} formed by 75 µl of extract (diluted 50-fold)

$$\equiv 0.030\,46 \times 50 \times 1000/75 \text{ µmol glyoxylate min}^{-1} \text{ formed by}$$
  1 ml undiluted extract
$$= \textbf{20.3 µmol glyoxylate min}^{-1} \textbf{ formed by 1 ml undiluted extract}$$

62 µg protein in 0.05 ml of extract diluted 10-fold
Therefore, $62 \times 10 \times 1 / 0.05$ µg protein in 1 ml undiluted extract = 12 400 µg
  protein

$$\equiv \textbf{12.4 mg protein in 1 ml undiluted extract}$$

Now we can determine that the specific activity of isocitrate lyase in the acetate-grown organisms = 20.3 / 12.4 µmol glyoxylate min^{-1} formed by 1 mg protein

Specific activity $\equiv$ 1.637 µmol glyoxylate min^{-1} (mg protein)$^{-1}$

The organisms now grow for 18.5 h on succinate, with a doubling time of 3.9 h

This means 18.5 / 3.9 generations will grow = 4.74 generations in 18.5 h

The increase in organisms will be $2^{4.74}$-fold in this time = 26.7-fold

Hence, **the final specific activity of isocitrate lyase after growth on succinate will be:**

$$1.637/26.7 = \textbf{0.0613 µmol glyoxylate min}^{-1}\textbf{(mg protein)}^{-1}$$

This first part of the question is long but really quite straightforward. The second part is hard (too hard for *all* the undergraduates that I've ever taught, 1st Class or otherwise). The calculation is in effect the same type as discussed in Chapter 8 (Enzymes) but is done in the opposite direction.

0.2 µmol acetate activated min^{-1} (mg protein)$^{-1}$ but only half (0.1 µmol) is used for biosynthesis

1 mol acetate (CH_3COO^-) contains 24 g carbon, and so 0.1 µmol acetate will contain 2.4 µg carbon

Carbon makes up 50% of the dry wt, so that 2.4 µg carbon will produce 4.8 µg organisms

Therefore the activity of acetate thiokinase is enough to make 4.8 µg organisms min^{-1} (mg protein)$^{-1}$

But 1 mg protein represents 2 mg organisms (because these contain 50% of their weight as protein).

Thus 1 mg organisms can make 2.4 µg new organisms min^{-1}

$$\equiv 0.0024 \text{ mg new organisms min}^{-1} \text{ (mg existing organisms)}^{-1}$$

The specific growth rate is therefore 0.0024 min^{-1}

$$0.0024 = \ln 2/t_d \text{ (t_d is the mean generation time)}$$

Therefore t_d = 0.693 / 0.0024 = 289 min

The maximum rate of growth (as a doubling time) would be 289 min

Problem 18

Bacteria in suspension

Absorbance of $1.0 \equiv 1.134$ mg dry wt $(\text{ml suspension})^{-1}$

$\therefore$ Absorbance of $\alpha \equiv 1.134 \, \alpha$ mg ml^{-1}

$\therefore$ Absorbance of $0.31 \equiv 1.134 \times 0.31$ mg ml$^{-1} = 0.3515$ mg ml^{-1}

Dilution is 5/40, so that undil. dry wt $= 0.3515 \times 40/5 = 2.812$ mg ml^{-1}

Hence 100 ml of culture contains 281.2 mg dry wt of organisms

Chlorophyll in 250 ml of extract

Absorbance of $1.054 \times 10^3 \equiv 10 \, \text{g L}^{-1}$ $(1\% \text{ w/v}) \equiv 10$ mg ml^{-1}

$\therefore$ Absorbance of $1 \equiv 10 \div (1.054 \times 10^3)$ mg ml^{-1}

$\therefore$ Absorbance of $0.5 \equiv 0.5 \times 10 \div (1.054 \times 10^3)$ mg $(\text{ml extract})^{-1}$

$\qquad\qquad = 4.74 \times 10^{-3}$ mg chlorophyll $(\text{ml extract})^{-1}$

$\therefore$ In 250 ml extract there are $4.74 \times 10^{-3} \times 250$ mg chlorophyll

$\qquad\qquad = \mathbf{1.185}$ **mg chlorophyll**

Moles of chlorophyll per g bacteria

250 ml of ether extract comes from the organisms in 100 ml of culture
Hence 281.2 mg bacteria contains 1.185 mg chlorophyll

$\therefore$ 1 g bacteria contains $1.185 \div 281.2$ g chlorophyll $= 4.21 \times 10^{-3}$ g

$\equiv 4.21$ mg $\equiv 4210$ µg chlorophyll g^{-1}

$\equiv 4210 \div 911 = \mathbf{4.62}$ **µmoles** $(\mathbf{g \ bacteria})^{-1}$

Chlorophyll content after 12 h

$12 \, \text{h} \equiv 12/5 = 2.4$ generations$(t_d = 5 \text{ h})$

$\therefore$ Increase in wt of bacteria $= 2^{2.4}$-fold $= 5.28$-fold

$\therefore$ Content of chlorophyll per unit wt of bacteria decreases 5.28-fold

$\equiv 4.62 \div 5.28 = \mathbf{0.875}$ **µmoles** $(\mathbf{g \ bacteria})^{-1}$

Problem 19

(1) The standard curve is shown in Fig. 17.7.

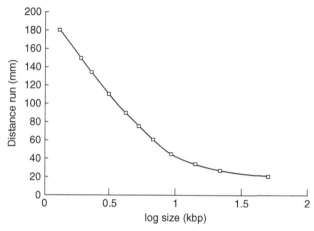

Fig. 17.7

(2) The restriction map for the vector is shown in Fig. 17.8 and that for the transducing phage in Fig. 17.9.

Fig. 17.8 Vector phage.

Fig. 17.9 Transducing phage.

Sizes of fragments (kbp)	Vector:	EcoRI	22; 11; 7
		HindIII	24; 16
		EcoRI + HindIII	22; 9; 7; 2
	Transducing phage:	EcoRI	22; 13; 7; 3
		HindIII	24; 16; 5
		EcoRI + HindIII	22; 9; 7; 4; 2; 1

(3) Vector 40 kbp; transducing phage 45 kbp; bacterial segment 5 kbp.

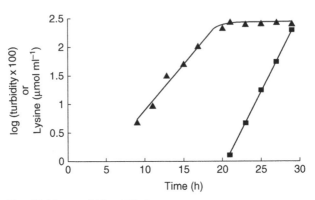

Fig. 17.10 ▲ turbidity; ■ lysine.

Problem 20

Note that by plotting log (turbidity × 100) one avoids having to plot negative values (see Fig. 17.10). Fortuitously (well, not really), the scale used for the logarithmic plot is also appropriate for the plot of concentrations of lysine as an arithmetic scale.

From the straight-line region of the graph of log turbidity against time the doubling time is 2 h. The fact that the plotted values are increased 100-fold does not alter the time needed for the exponential increase to advance by 0.301 ($= \log 2$).

(1) Excretion of lysine occurs at a constant (non-exponential) rate: 2.21 µmol lysine ml^{-1} are excreted in the period between 21 h and 29 h

$$\equiv 276 \text{ nmol } ml^{-1}h^{-1}$$
$$\equiv 4.6 \text{ nmol } ml^{-1}min^{-1}$$

By plotting growth and lysine excretion on the same graph it is obvious that lysine appears in the medium only when growth has stopped and the bacterial population is constant.

While this excretion is occurring the bacterial dry wt is
$2.8 \times 0.35 \text{ mg } ml^{-1} = 0.98 \text{ mg } ml^{-1}$
Hence $4.6 / 0.98 = 4.7$ nmol lysine excreted min^{-1} by 1 mg bacteria
Therefore 9.4 nmol excreted min^{-1} by 1 mg protein
The observed enzymic activity is just adequate to account for the rate of excretion.

(2) When doubling time is 120 min then $\mu = 0.693 / 120 = 0.005\,775$ min^{-1}
i.e. 1 mg existing bacteria makes 0.005 775 mg new bacteria min^{-1}
So that 0.5 mg protein makes 0.005 775 mg new bacteria min^{-1}
1 mg protein makes $2 \times 0.005\,775$ mg new bacteria min^{-1}

$$\equiv 11.55 \text{ µg new bacteria min}^{-1} \text{ (mg protein)}^{-1}$$

5% of bacterial wt is lysine $= 0.5775$ µg lysine made min^{-1} (mg protein)$^{-1}$
577.5 ng lysine $= 577.5 / 146$ nmoles min^{-1} (mg protein)$^{-1}$

$$= 3.96 \text{ nmoles min}^{-1} \text{ (mg protein)}^{-1}$$

The observed enzymic activity is adequate for the rate of growth.

Problem 21

(1) *Eco*RI produces two fragments from the circular plasmid, and so there
must be two *Eco*RI sites, separated by 2.5 kbp. This 2.5 kbp fragment is
not attacked by *Eco*RI and must be the *Eco*RI fragment of *B. subtilis*
DNA that is inserted into plasmid pAB2. It has been inserted at the
*Eco*RI site at the cloning site. Consequently the *Hind*III site of the
cloning site will be separated by 2.5 kbp from its original position (see
Fig. 17.11).

HindIII produces three fragments, so that there must be three sites
for its action. One of these is immediately to the right of the *Eco*RI site
that is no longer in its original position at the cloning site. The other two

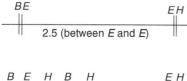

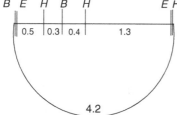

Fig. 17.11 Restriction map of plasmid pAB2.

sites must be within the inserted 2.5 kbp because there is only one HindIII site in the pAB1 plasmid.

BamHI produces two fragments, and so must have two sites of action, one being in the cloning site immediately to the left of the EcoRI site (in the cloning region). The second site must be to the right of the EcoRI site (which is still in the original cloning site), within the inserted DNA, also the combined action of BamHI and HindIII could not otherwise yield a fragment as large as 4.2 kbp. BamHI cuts the 0.7 kbp fragment (that HindIII acting alone produces) into 0.3 and 0.4 kbp fragments. Hence the position of the HindIII 0.7 kbp fragment is defined.

(2) 6.02×10^{23} plasmids weigh $300 \times 2 \times 6.7 \times 10^3$ g

$= 6 \times 6.7 \times 10^5$ g $= 4.02 \times 10^6$ g

$(6.02 \times 10^{23})/(4.02 \times 10^6)$ plasmids weigh 1 g

That is, 1.5×10^{17} plasmids weigh 1 g

1.5×10^{11} plasmids weigh 1 μg

3×10^{12} plasmids weigh 20 μg

Therefore the organisms from 1 ml of culture contain 3×10^{10} plasmids
200 colonies from 0.1 ml of a 1×10^{-5} dilution of the culture
$\therefore 2 \times 10^8$ colonies from 1 ml of undiluted culture
3×10^{10} plasmids from 2×10^8 bacteria
$= 3 \times 10^{10}/2 \times 10^8$ plasmids per bacterium
$= 1.5 \times 10^2$
$= \textbf{150 plasmids per bacterium}$

Problem 22

(1) 1 mM thymine in 0.1 M HCl has optical density $= 8.0 \equiv 1$ μmol ml^{-1}
Therefore OD $1 \equiv 1 / 8$ μmol ml^{-1}
and OD $x \equiv x / 8$ μmol ml^{-1}
Hence, OD $0.28 \equiv 0.28 / 8$ μmol ml^{-1}
And so 5 ml 0.1 M HCl contain $5 \times 0.28 / 8$ μmol thymine

The thymine spot on the chromatogram contained 0.175 μmol
By similar calculations we find that:

The adenine spot contained 0.175 μmol
The cytosine spot contained 0.075 μmol
The guanine spot contained 0.077 μmol

A + T + G + C = 0.502 μmol and G + C = 0.152 μmol

$$\%GC = 100 \times 0.152/0.502 = 30\%$$

(2) 1 mole thymine can produce 1 mole thymidylic acid (TMP)

    1 mole TMP + 1 mole dAMP

    = 1 mole AT base pairs = 614 g

    Because %GC = 30, 7 moles AT base pairs occur in every 10 moles of
      base pairs in DNA

    Thus 7 moles of thymine can produce 10 × 614 g DNA

    7 × 126 g thymine can make 6140 g DNA

    882 / 6140 g thymine can make 1 g DNA

    0.144 g thymine makes 1 g DNA

(3) 144 mg thymine makes 1 g DNA

    1 g bacteria contains 50 mg DNA

    Hence 144 / 20 mg thymine is needed to make the DNA in 1 g bacteria

    7.2 mg thymine is needed to grow 1 g bacteria.

    (1 L of medium is needed to grow 1 g bacteria, and in 1 L there is 10 mg
  thymine)

(4) 0.144 g thymine makes 1 g DNA

    so that 144 μg thymine makes 1 mg DNA

    Hence 144 μg thymine must be labelled with at least 1×10^6 cpm

    If these counts come from ^{3}H (20% efficiency of counting) then at least
  5×10^6 dpm must be associated with 144 μg thymine

    $5 \times 10^6 / 144$ dpm must represent 1 μg thymine

    In 1 L of the medium there is 10 000 μg thymine, so that $10^4 \times 5 \times 10^6 /$
  144 dpm will be needed

$$5 \times 10^{10}/144 = 3.47 \times 10^8 \text{ dpm} \equiv 3.47 \times 10^8/2.2 \times 10^6 \text{ μCi}$$

$$= 158 \text{ μCi } [^3\text{H}]\text{-thymine L}^{-1}$$

^{14}C is counted four times more efficiently than ^{3}H, so that 158 / 4 μCi
would suffice:

$$= 40(39.5) \text{ } \mu\text{Ci} \text{ } [^{14}\text{C}]\text{-thymine L}^{-1}$$

(5) To isolate 10 mg of purified labelled DNA it is necessary to start with 100 mg DNA. This quantity will be present in 2 g bacteria, and so to get this weight of organisms, 2 L of medium will be required.

Therefore, either 316 μCi [³H]-thymine or 80 μCi [¹⁴C]-thymine will be needed.

The [³H]-thymine would cost £137, whereas 100 μCi [¹⁴C]-thymine could be bought for £112.

The best buy is two packages each of 50 μCi [¹⁴C]-thymine.

(6) Yes, it does matter. If %GC were 25, then %AT would be 75. Slightly more thymine would be present per gram of DNA than has been calculated above, and slightly less radiolabel would be needed to get the same specific activity in the DNA. If the %GC were 75, then %AT would be 25. In this case there would be less thymine (70 / 25 times less) per gram of DNA. Nearly three times as much labelled thymine would be needed to achieve the desired specific activity in the DNA. The best buy would be 1 mCi of [³H]-thymine (approx. 0.9 mCi used) for £137. If [¹⁴C]-thymine were still chosen, 250 μCi (approx. 240 μCi used) would cost £183.

Problem 23

(1) Crude extract

Minus substrate absorbance changed by 0.42 in 7 min $\equiv 0.06 \text{ min}^{-1} =$ blank rate

With substrate absorbance changed by 0.60 in 3 min $\equiv 0.20 \text{ min}^{-1}$

Correcting for blank rate, enzymic rate is 0.14 min^{-1}

Absorbance of $6.22 \times 10^{3} = 1$ molar $= 1 \text{ mol L}^{-1} \equiv 1 \text{ mmol ml}^{-1}$

Absorbance of $6.22 \equiv 1 \text{ } \mu\text{mol ml}^{-1}$

Absorbance of $1 \equiv 1 \text{ / } 6.22 \text{ } \mu\text{mol ml}^{-1} = 3 \text{ / } 6.22 \text{ } \mu\text{mol per 3 ml of}$ cuvette contents

Absorbance of $x = 3x \text{ / } 6.22 \text{ } \mu\text{mol per 3 ml of cuvette contents}$

Hence absorbance change of $0.14 \text{ min}^{-1} = 3 \times 0.14 \text{ / } 6.22 = 0.0675 \text{ } \mu\text{mol} \text{ min}^{-1}$

This was catalysed by 10 μl of diluted extract, which contained 20 μg protein

Therefore **specific activity is 3.375 μmol min⁻¹ (mg protein)⁻¹**

(2) Purified enzyme

Minus substrate absorbance changed by 0.12 in 6 min $\equiv$ 0.02 min^{-1} = blank rate

With substrate absorbance changed by 0.45 in 2 min $\equiv$ 0.225 min^{-1}

Correcting for blank rate, enzymic rate is 0.205 min^{-1}

Absorbance change of 0.205 min^{-1} = 3 × 0.205 / 6.22 = 0.0989 μmol min^{-1}

This was catalysed by 30 μl of diluted enzyme

And so 10 μl of diluted enzyme (1.5 μg protein) would catalyse 0.032 96 μmol min^{-1}

Therefore **specific activity is 22.0 μmol min^{-1} (mg protein)$^{-1}$**

(3) **Purification factor is 22 / 3.38 = 6.5-fold**

Problem 24

(1) We start by assuming that the UV spectrum of the dipeptide is similar to that of uridine, though it is unlikely to be extremely similar owing to the absorption at 280 nm that is due to the peptide bonds. The 280 : 260 ratio is 0.47 which is fairly close to that of uridine diphosphate.

With uridine, an absorption of $1 \times 10^4 \equiv 1$ M

Hence absorption of $1 \equiv 0.1$ mM

And $0.53 \equiv 0.053$ mM

This absorption was given by a 1 / 100 dilution of the solution of dipeptide, so that undiluted the solution of dipeptide is about 5 mM (based on the extinction coefficient of uridine diphosphate).

1 ml of the solution of peptide contained 11 μmol of phosphorus $\equiv$ 11 mM.

This is consistent with the above estimate, because two atoms of P should be present for one molecule of uridine.

1 ml of hydrolysate contains 0.88 μmol hexosamine

This amount was originally present in 0.2 ml of the dipeptide solution

So that 1 ml of peptide solution contained 4.4 μmol hexosamine $\equiv$ 4.4 mM.

This is again consistent with the above results – there should be one molecule of hexosamine accompanying one molecule of uridine diphosphate. Some destruction of hexosamine is likely to occur during acid hydrolysis of the dipeptide.

1 ml of hydrolysate contains 1.05 μmol acetate.

This amount was originally present in 0.2 ml of the dipeptide solution

So that 1 ml of peptide solution contained 5.25 μmol acetate ≡ 5.3 mM. This is consistent with the earlier results – one molecule of muramic acid in the dipeptide should be mono-acetylated.

Chromatography showed that alanine and glutamate (and no other amino acids) were present in the presumed dipeptide. The lengthy hydrolysis with strong acid would destroy most of the muramic acid that would have been released from the dipeptide, so that no spot from muramic acid would be expected on the chromatogram.

Thus, the analyses are consistent with the structure of the dipeptide.

The assays for phosphorus and for acetate are the least subject to doubt, and so the concentration of the solution of dipeptide is

$$(5.5 + 5.3)/2\,\text{mM} = \textbf{5.4 mM}$$

(2) Ideally there should be no more radioactivity at the origin in the 'minus dipeptide' or in the 'minus enzyme' controls than in the 'minus [^{3}H]-Dap' control. The crude extract might contain some ATP, which could explain why the 'minus ATP' control is relatively high. There are some counts at the origin even when no radioactivity was added to the assay mixture. These counts must represent background.

If this background is subtracted from the above controls then the controls are very small (22, 14, 53 dpm) in relation to the counts from the complete system (31, 126, 229, 352 dpm).

Take the average of the first two controls (18 dpm), then use this value to correct the results from the complete system:

	Time (min)	dpm from origin of chromatogram
Complete system	0	13
Complete system	5	108
Complete system	10	211
Complete system	15	334

Now we need to relate dpm at the origin to amounts (nmoles) of diaminopimelate.

5 mM diaminopimelate ≡ 5 mmol L^{-1} ≡ 5 μmol ml^{-1} ≡ 5 nmol μl^{-1}

Hence in 4 µl of 5 mM diaminopimelate there are 20 nmoles

In 1 ml of [^{3}H]-Dap there are 1×10^6 dpm $\equiv 1 \times 10^3$ dpm µl^{-1}

Hence in 4 µl of [^{3}H]-Dap there are 4×10^3 dpm

This means that 4×10^3 dpm represent 20 nmol diaminopimelate, so that 1 dpm represents 0.005 nmol diaminopimelate.

Counts at the origin can be expressed as amounts of diaminopimelate bound:

0 min 13×0.005 nmol = 0.065 nmol

5 min 108×0.005 nmol = 0.54 nmol

10 min 211×0.005 nmol = 1.06 nmol

15 min 334×0.005 nmol = 1.67 nmol

These data show an obvious linear increase with time.

Therefore we can say that in 15 min approximately 1.6 nmol is bound.

This amount came from only 40% of the total assay system, so that in the complete system (50 µl) 4 nmoles was bound in 15 min.

In 1 min 0.27 nmol was bound.

In the complete system (50 µl) there is 20 µg protein (work it out).

Consequently 0.27 nmol bound per min by 20 µg protein

$\equiv$ 270 nmol min^{-1} bound by 20 mg protein

$\equiv$ (13.5) **14 nmole diaminopimelate bound min^{-1} (mg protein)$^{-1}$**

$\equiv$ **0.014 units of Dap-adding enzyme (mg protein)$^{-1}$**

Problem 25

The volume of organisms plus capsule ($r = 1$ µm) is $4\pi \times 1^3 / 3$ µm^3 = $4\pi / 3$ µm^3

Volume of organism alone ($r = 0.5$ µm) = $4\pi \times 0.5^3 / 3$ µm^3 = $4\pi \times 0.125 / 3$ µm^3

Hence, volume of capsule = $(4\pi/3) - (0.5\pi/3) = 3.5\pi/3$µm^3

Half of this volume is water, so the volume of polysaccharide is $1.75\pi / 3$ µm^3

1 ml of polysaccharide weighs 1.00 g

1 ml = $(1 \times 10^{-2})^3$ m^3 = 1×10^{-6} m^3

1 µm^3 = $(1 \times 10^{-6})^3$ m^3 = 1×10^{-18} m^3

$\therefore$ 1 µm^3 = 1×10^{-12} ml

So that 1 µm^3 of polysaccharide weighs 1×10^{-12} g

$\therefore 1.75\pi / 3 \ \mu m^3$ of polysaccharide weighs $(1.75\pi / 3) \times 1 \times 10^{-12} g$
$= 1.83 \times 10^{-12} \ g$

One capsule contains 1.83×10^{-12} g of polysaccharide

500 000 g polysaccharide represents 6×10^{23} molecules and so 6×10^{23}
terminal units of glucose
$\therefore$ 1 g polysaccharide has 1.2×10^{18} terminal glucose units
And 1×10^{-12} g polysaccharide has 1.2×10^6 terminal glucose units
1.83×10^{-12} g polysaccharide has $1.83 \times 1.2 \times 10^6$ terminal glucose units

$= \mathbf{2.20 \times 10^6}$ **terminal glucose units on one capsule**

The surface area of the capsule $(r = 1 \ \mu m)$ is $4\pi \times 1^2 \ \mu m^2$
Thus, 2.20×10^6 terminal glucose units are spread over an area of $4\pi \ \mu m^2$
$\equiv (2.20 \times 10^6) / (4\pi)$ terminal glucose units per μm^2

$= \mathbf{1.75 \times 10^5}$ **terminal glucose units per μm^2 of capsular surface**

Problem 26

(1) See Fig. 17.12. With *Eco*RI four fragments or fewer: there will be four
unless the two fragments (from the second insertion) happen to be of
equal size, or either is equal in size to one of the fragments from the first
insertion site. In these cases there would be three hybridising bands.
There will be only two bands if the two fragments from insertion 2 were
equal in size and equal to one of the bands from site 1, or if each one
fragment from insertion 2 is equal to one fragment from site 1. With
*Hind*III four or fewer: the same argument applies here.

(2) See Fig. 17.13.

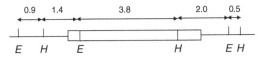

Fig. 17.12 Transposon insertion.

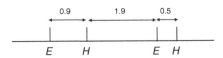

Fig. 17.13 Wild-type chromosome.

(3) With both five or fewer: if every fragment that includes chromosomal DNA is of different size, and none of these is of size 3.8 kbp there will be five bands. At each insertion site the central region of the transposon will yield a 3.8 kbp fragment.

Problem 27

(1) See Fig. 17.14. From this curve one can read off that 1 ml of the spore extract contained 470 µg DPA

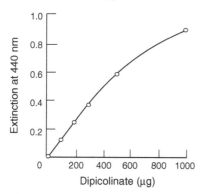

Fig. 17.14

Hence 5 ml extract contained 2350 µg DPA

This means that 50 mg spores contained 2.35 mg DPA

DPA is a dibasic acid, and so will lose 2 H^+ ions when one molecule of DPA is converted to its Ca^{2+} salt. The mol. wt of Ca dipicolinate is 167 − 2 + 40 = 205 Da

And 50 mg spores contain 2.35 × 205 / 167 mg calcium dipicolinate = 2.88 mg

Therefore calcium dipicolinate is 2 × 2.88 g per 100 g spores

$$= 5.8\%(w/w) \text{ of these spores}$$

(2) The extinction was 0.234 when 200 µg DPA was in 5 ml of the assay mixture

That is, 200 / 167 µmol = 1.20 µmol in 5 ml

And so there will be 1.20 µmol DPA-Fe^{2+} complex in 5 ml ≡ 0.24 µmol ml^{-1}

Thus a 0.24 mM solution of DPA-Fe^{2+} has an extinction of 0.234

and a 0.24 M solution will have an extinction of 234

and a 1 M solution will have an extinction of 234 / 0.24 = 975

The extinction coefficient of DPA-Fe^{2+} complex is 9.75 $\times$ 10^2 L mol^{-1} cm^{-1}

(3) 1 M Ca DPA has an extinction of 5.4 $\times$ 10^3

∴ 1 mM Ca DPA has an extinction of 5.4

and so extinction of 5.4 $\equiv$ 1 μmol ml^{-1}

and extinction of 1 $\equiv$ 1 / 5.4 μmol ml^{-1}; extinction of 0.154 $\equiv$ 0.154 / 5.4 μmol ml^{-1}

The undiluted supernatant liquid contains 5 $\times$ 0.154 / 5.4 μmol ml^{-1}

So, 4 ml of undiluted supernatant liquid contains 4 $\times$ 5 $\times$ 0.154 / 5.4 μmol Ca DPA

$$\equiv 205 \times 4 \times 5 \times 0.154/5.4 \,\mu g \text{ Ca DPA} = 117 \,\mu g$$

This amount came from 2 mg spores

∴ 1000 μg spores contain 58.5 μg Ca DPA

Ca DPA is 5.9% (w / w) of the spore

The two methods give results in agreement.

Problem 28

(1) The plot (not shown) gives a straight line which intercepts the y axis (μ [= D]) at the value of q_m **(for oxygen). This reads off as 1.7 mmol g^{-1} h^{-1}.**

To determine Y_g for oxygen we use the relations:

$$1/Y = 1/Y_g + q_m/\mu = q/\mu$$

To get Y_g into the required units (g mol^{-1}) we must express q and q_m in moles

So, taking μ as 0.1 we can write $1/Y = 0.0033 / 0.1 = 0.033$

Thus $1/Y_g + 0.0017 / 0.1 = 0.033$

$1/Y_g = 0.033 - 0.017 = 0.016$

$$Y_g = \textbf{62 g (mol O}_2\textbf{)}^{-1}$$

The value of Y_g comes out close to this number no matter which value of μ (and the corresponding value of q) is taken.

(2) The plot (not shown) gives a straight line which intercepts the y axis $(1/\mu$ $[=1/D])$ at the value of $1/\mu_{max}$ and intercepts the x axis $(1/s)$ at $-1/K_s$.

So, $1/\mu_{max} = 0.6$ and therefore $\mu_{max} = 1.6\ h^{-1}$

And $-1/K_s = -10$ hence $K_s = 0.1\ mmol\ L^{-1}$

(3) Concentrations of glucose $(mmol\ L^{-1})$ are $2.2/180$ and $5.1/180$ so that we can write

$$0.0122(\mu_{max}-0.3)/0.3 = 0.0283(\mu_{max}-0.4)/0.4,\ \textbf{and solve for}\ \mu_{max}$$

Cross-multiply to get

$0.004\ 88(\mu_{max} - 0.3) = 0.008\ 49(\mu_{max} - 0.4)$

$\mu_{max} - 0.3 = 0.008\ 49(\mu_{max} - 0.4)\ /\ 0.00488$

$\mu_{max} - 0.3 = 1.74(\mu_{max} - 0.4)$

$\mu_{max} - 0.3 = 1.74\mu_{max} - 0.696$

$\mu_{max} = 1.74\mu_{max} - 0.696 + 0.3$

$\mu_{max} = 1.74\mu_{max} - 0.396$

Hence, $0.396 = 0.74\mu_{max}$

$\mu_{max} = 0.396\ /\ 0.74 = 0.535\ h^{-1}$

Now we can find K_s (twice to check that μ_{max} seems to be correct)

$$K_s = 0.0122(0.535 - 0.3)/0.3$$
$$= 0.0122 \times 0.235/0.3 = 9.56 \times 10^{-3}\ mmol\ L^{-1}$$
$$K_s = 0.0283(0.535 - 0.4)/0.4$$
$$= 0.0283 \times 0.135/0.4 = 9.55 \times 10^{-3}\ mmol\ L^{-1}$$

Problem 29

(1) Requirement for aromatic amino acids:

1 mg bacteria $\equiv$ 500 μg protein, containing $500 \times 15/100 = 75$ μg aromatics

$\equiv 75/165$ μmoles $= 0.455$ μmoles $\equiv \textbf{0.455} \times \textbf{10}^{-3}\,\textbf{mmoles mg}^{-1}$

(2) 10 nmoles of erythrose 4-phosphate made min^{-1} $(mg\ protein)^{-1}$ $(\equiv 2\ mg$ bacteria)

10 nmoles erythrose 4-phosphate makes 10 nmoles of aromatic amino acids

$\equiv 10 \times 165$ ng $\equiv 1.65$ μg

1.65 μg aromatic amino acids can form $1.65 \times 100/15 = 11$ μg protein

11 µg protein can yield 22 µg bacteria (protein is 50% w / w)

Thus, 2 mg bacteria can make not more than 22 µg new organisms min^{-1} (if transketolase is rate-limiting)

1 mg existing bacteria can make 0.011 mg new organisms min^{-1}

Hence, maximum specific growth rate = **0.011 min^{-1}**

This is equivalent to a t_d of 0.693 / 0.011 = 63 min. The observed t_d was 120 min which is a slower rate than that which transketolase is able to maintain. Therefore, transketolase is **not** rate-limiting (unless the activity inside the organisms is less than was measured with an extract under optimal conditions of assay).

Problem 30

(1) Fig. 17.15.
(2) Fig. 17.16.

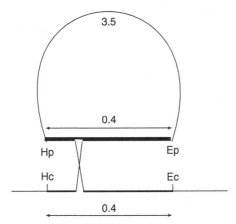

Fig. 17.15 Crossover.

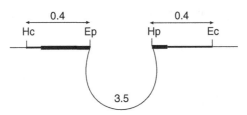

Fig. 17.16 Inserted plasmid.

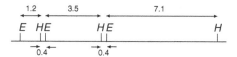

Fig. 17.17 Result of crossover.

Fig. 17.18 Wild-type chromosome.

(3) Fig. 17.17; region for insert (see Fig. 17.18):

EcoRI	1.6 kbp
HindIII	7.5 kbp
EcoRI + HindIII	0.4 kbp

Problem 31

(1) 38.6 kcal −1.68 V + 0.6 V = −1.08 V
(2) 33.6 kcal −1.46 V + 0.8 V = −0.66 V
(3) 500 candela deliver 1×10^{17} quanta s^{-1} cm^{-2}, which is equivalent to 3.6×10^{20} h^{-1} cm^{-2}

Hence, the whole culture receives 3.6×10^{22} quanta h^{-1}
Only 10% of these are available for photosynthesis = 3.6×10^{21} quanta h^{-1}
This represents $40 \times 3.6 \times 10^{21}$ / (6.02×10^{23}) kcal h^{-1}

   = 0.239 kcal h^{-1}

This is the maximum amount of energy that a very dense culture might absorb. Less of the available energy will be taken up by a sparse population.

Problem 32

(1) After correcting for background, 0.1 μg of labelled methionine gave 3552 cpm

At 82% efficiency, this represents 3552 × 100 / 82 dpm = 4332 dpm

4332 dpm ≡ 4332 / 60 dps = 72.2 dps

Hence 722 dps represent 1 µg methionine

and 722 000 dps represent 1 mg methionine ≡ 722 kBq

(2) After 10 days the activity will decrease by a factor of 1 / $2^{(10 / 87.1)}$

$$2^{(10/87.1)} = 2^{0.1148} = 1.0828$$

The activity will be 722 / 1.0828 = 667 kBq (mg methionine)$^{-1}$

(3) In the translation mix (1 ml) there are 50 µmol methionine

So that 667 kBq represent 50 µmol methionine

and 667 Bq (= 667 dps) represent 50 nmol methionine

Thus 1 dps represents 50 / 667 nmol methionine

537 dps represent 50 × 537 / 667 = 40.25 nmol methionine

So 0.3 mg protein contains 40.25 nmol methionine

$$\equiv 40.25 \, / \, 0.3 \, \text{nmol mg}^{-1} = 134 \, \text{nmol mg}^{-1}$$

$$= 0.134 \, \mu\text{mol methionine (mg protein)}^{-1}$$

There is 134 µmol methionine (g protein)$^{-1}$

$$14\,925 \text{ g protein } (1 \text{ mol}) \text{ will contain } 14\,925 \times 134 \mu\text{mol}$$
$$\text{methionine} = 1\,999\,950 \, \mu\text{mol}$$

1 mol of protein contains 1 999 950 / 1 × 10^6 mol methionine = 2 mol methionine

2 residues of methionine are present in one molecule of protein

Problem 33

(1) 1 mM NADPH has extinction of 6.22 ≡ 1 mmol L^{-1} ≡ 1 µmol ml^{-1}

An extinction of 1 = 1 / 6.22 µmol ml^{-1}

An extinction of $x = x$ / 6.22 µmol ml^{-1}

In 3 ml, an extinction of $x = 3x$ / 6.22 µmol ml^{-1}

At pH 8.0 the dehydrogenase (0.5 ml) gives an increase of extinction of 0.35 / min

$\equiv$ 3 × 0.35 / 6.22 µmol formed min^{-1} = 0.169 µmol min^{-1} at pH 8.0

At pH 10.5, 0.5 unit of dehydrogenase would form 0.5 µmol min^{-1}

∴ **Ratio of activities at pH 8.0 and pH 10.5 is 0.169 : 0.5 = 0.34 : 1.00**

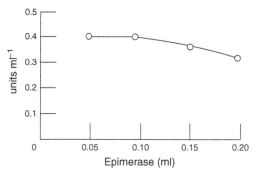

Fig. 17.19

(2) Fig. 17.19. From the graph the best estimate of the epimerase activity is 0.4 units ml^{-1}

$$= 0.05 \text{ units (mg protein)}^{-1}$$

(3) 0.5 unit of dehydrogenase is sufficient for up to 0.04 unit of epimerase
∴ **To assay 0.1 unit of epimerase 1.25 units of dehydrogenase are needed**

Problem 34: One more for fun

(1) What *can't* be Smith's age?

It can't be a prime number. Suppose that his age is n, a prime number, then the only possible way for three ages to give this product would be $n.1.1$. Even if this possibility were accepted – Smith would have to be 0 when last he saw Jones – it would mean that Smith would know the ages without needing more information.

It can't be a number with only one set of three factors, otherwise Smith would be sure of the ages without the second or last piece of information (biting nails). For example, 25 can only be 5.5.1 (or 25.1.1); 34 can only be 17.2.1 or (34.1.1).

(2) It follows that Smith's age must have more than one set of three factors, in addition to $n.1.1$.

(3) It then becomes clear that two of the sets of the three factors of Smith's age must add to the same sum (his room number) otherwise the last piece of information would be unnecessary in finding the ages of the children.

(4) Some work now has to be done in order to seek out what numbers satisfy point 3.

If we start looking between 30 and 50 (as guesses at Smith's age) we find that:

36 can be 6.6.1 and 9.2.2, both of which add to 13;

40 can be 8.5.1 and 10.2.2, both of which add to 14.

There are no other possibilities between 30 and 50. (There are no other possibilities between 0 and 120.)

(5) If Smith's age were 40, then the last piece of information (biting nails) would not help in deciding the ages of the children. However, this last piece of information does make 9.2.2 more likely than 6.6.1. The choice isn't entirely conclusive (one of the 6's could be older than the other; they don't have to be twins) but Smith's final words suggest that he is not entirely certain, though he does give the correct answer.

Consequently, Smith is 36, his room number is 13, and **the ages are 9, 2, 2**.

Conclusion

As a teacher of some experience, the writer is well aware of the great value to be attached to the solution of numerical problems as an aid to the understanding of the principles of physical chemistry.

Samuel Glasstone

What's the point of working through these diabolical questions? What's the point of making up these questions? Are they just 'a madman's fly trap', as the plots of a mystery writer (John Dickson Carr) that I much enjoy have been described? Obviously, I think there are good points, otherwise I wouldn't be writing this book, and would not be keeping myself sequestered nor neglecting important household tasks.

The first point is that to become a scientist one does need to develop awareness of the ways in which experimental data are turned into conclusions. How was Avogadro's number evaluated? How did Millikan (of the oil-drop experiment) determine the size of the charge on an electron? These were things I had to write about as an undergraduate, and I had to do the calculations too. Neither then nor now did I see the great relevance of this to biochemistry, but I did begin to see how hard-won are scientific 'facts', and I certainly realised then that I could manage handling the numbers, and using such horrors as five-figure logarithmic tables, long before the arrival of calculators.

This leads to the next point – doing problems helps to improve one's mathematical skill, and gives the confidence to believe that one has such skill. Miss Dollan taught me how to do long division of pounds, shillings and pence, which at the time was for me just as hard as any of the maths in this book. From her I gained the assurance that I could cope, and I don't forget Miss Dollan.

For an author, making up data-handling problems is an interesting challenge. The writing must be unambiguous, the problem a bit different every time, and the solution clear but not too easy nor too hard to reach.

Real experiments do not, of course, always give rise to clear solutions. It may be surprising, but all the problems given here are not mad fabrications, but derive from actual experiences in the laboratory. Even the one about the number of glucose residues on the staphylococcal surface, which looks very contrived, came from a request for such a calculation by a colleague in another university.

So far, data-handling has been viewed as a learning process, and so it ought to be. Attendance at a university is also meant as a learning process, but we do impose examinations at various stages. Data-handling ability gets examined because it is a skill that a graduate should have acquired. It is unsatisfactory to put a data-handling problem on an examination paper as one question among several others that all require essays as answers, especially when there is a choice, and the problem is not compulsory. Many candidates will evade the problem, while those who do it (as I always would in my days as an undergraduate) may get a very high mark (or a very low one) outside the range of marks that are generally given for essays.

Having an examination which is made up entirely of data-handling questions is best. It should be held at the end of the final year, when the students have gained some knowledge. Thirty years ago, we began by setting four long questions, any three of which a student was required to answer, with four hours as a supposedly generous allowance of time. Our experience from several years with this format was that no students scored outstandingly highly over the whole paper, but a few did very badly – on a long intricate question it is possible to get completely lost. More recently some shorter problems have been included. These are not *necessarily* easier, but they are meant to give the less talented students the chance to show that they are not without ability in this area. Do you call this dumbing down? I'm not sure what I think.

Computers cannot be used in a written examination, basically because they can be preprogrammed with too much helpful information. In any case, using a computer is not very profitable when one has to do only a small number of calculations that are of a novel kind. The effort of writing and debugging a program is then disproportionate and not worthwhile. It's a very different story when there are large numbers of repetitive calculations to make. As one who spent all of Coronation Day (1953, not 1937) laboriously working out results from practical classes in which I was supposed to learn Warburg manometry, I can thoroughly appreciate the value of a spreadsheet. Indeed

I believe that every scientist ought to be able to construct a spreadsheet program. It's a lot harder to write programs in Pascal or C.

The last words are that data-handling gets easier the more you do it. Practice, practice, practice. Get confident.

I flatter myself that from the long experience I have had, and the unceasing assiduity with which I have pursued these studies in which you and I have been engaged, I shall be acquitted of vanity in offering some hints to your consideration. They are indeed in a great degree founded upon my own mistakes in the same pursuit ...

Sir Joshua Reynolds

Further reading

There are few books that discuss how to tackle data-handling in this area of science.

The best is *Quantitative Problems in Biochemistry*, by E. A. Dawes (Edinburgh: Churchill Livingstone). This went through several editions from the 1960s up to the 1980s, but has been long out of print. However, many libraries will hold one or more copies.

Practical Skills in Biomolecular Sciences, by R. Reed, D. Holmes, J. Weyers and A. Jones (3rd edn, 2007, Cambridge: Pearson Publishing), is so full of excellent advice to undergraduates that it is indispensable. The section about data-handling is brief, but good.

Maths from Scratch for Biologists, by A. J. Cann (Chichester: John Wiley), published in paperback (2003), is a very helpful book that is in print.

Textbooks of biochemistry or microbiology contain many short problems, though there is rarely much explanation of how to do them. Elementary texts on statistics, with abundant problems and detailed solutions, are very easy to find in shops or libraries. One of the best is *Statistics: A First Course*, by J. E. Freund and G. A. Simon (5th edn, 1991, Englewood Cliffs, NJ: Prentice Hall).

Index

Printed in the United States
by Baker & Taylor Publisher Services